VI International Forum for Young Scientists "Space Engineering 2018"

MATEC Web of Conferences Volume 158 (2018)

Tomsk, Russia
26-28 April 2018

Editors:

V. Borikov
S. Uchaikin
P. Baranov
V. Ivanova
A. Dolgih
I. Minin
O. Minin

ISBN: 978-1-5108-5978-4

Printed from e-media with permission by:

Curran Associates, Inc.
57 Morehouse Lane
Red Hook, NY 12571

CURRAN ASSOCIATES INC. proceedings.com

Some format issues inherent in the e-media version may also appear in this print version.

Printed by Curran Associates, Inc. (2018)

For additional information, please contact EDP Sciences – Web of Conferences at the address below.

EDP Sciences – Web of Conferences
17, Avenue du Hoggar
Parc d'Activité de Courtabœuf
BP 112
F-91944 Les Ulis Cedex A
France

Phone: +33 (0) 1 69 18 75 75
Fax: +33 (0) 1 69 28 84 91

contact-edps@webofconferences.org

Additional copies of this publication are available from:

Curran Associates, Inc.
57 Morehouse Lane
Red Hook, NY 12571 USA
Phone: 845-758-0400
Fax: 845-758-2633
Email: curran@proceedings.com
Web: www.proceedings.com

TABLE OF CONTENTS

*** Formatting Issue - Best Available Paper/Graphic ***

Statement of Peer review

In submitting conference proceedings to *Web of Conferences*, the editors of the proceedings certify to the Publisher that

1. They adhere to its **Policy on Publishing Integrity** in order to safeguard good scientific practice in publishing.
2. All articles have been subjected to peer review administered by the proceedings editors.
3. Reviews have been conducted by expert referees, who have been requested to provide unbiased and constructive comments aimed, whenever possible, at improving the work.
4. Proceedings editors have taken all reasonable steps to ensure the quality of the materials they publish and their decision to accept or reject a paper for publication has been based only on the merits of the work and the relevance to the journal.

Title, date and place of the conference

[illegible]

[illegible] Tomsk, Russia

Proceedings editor(s):

[illegible]

Date and editor's signature

[illegible]

17, avenue du Hoggar - PA de Courtabœuf – BP 112 - 91944 Les Ulis Cedex A (France)
Tél. : 33 (0)1 69 18 75 75 - Fax : 33(0)1 69 07 45 17 – www.edpsciences.org

There is need in new generation smart grid for the space and ground energy systems

Alexander Ageev[1], *Svetlana* Bortalevich[2], *Evgeny* Loginov[2,*], *Alexander* Shkuta[3], and *Dmitry* Sorokin[2]

[1]RAS, Institute of Economic Strategies, 101000 Moscow, Russia
[2]RAS, Institute of Market Problems, 117418 Moscow, Russia
[3]Financial University under the Government of the Russian Federation, World Economics and International Business Department, 125993 Moscow, Russia

Abstract. The aim of the article is to considerate the opportunities of synchronization of the space and ground systems that generate and transfer energy on the basis of new generation smart grid using. The authors substantiate the necessity of new intellectual monitoring services that assess the processes took place in "generation-transportation-distribution-consumption" space and ground systems. This is made in order to improve the dynamic indicators of the energy system and to avoid the emergencies. The authors also give a prognosis of the dynamic indicators of the electric power super-system in analyzing metastable conditions in different energy modes.

1 Introduction

The problem of the space and ground systems integration and electric power transfer as part of integrated super-system is high relevant nowadays. Especially it is because of the projects of the energy sources transferring from the Earth to space [1].

However, this problem may be solved very soon due to the modern technological progress.

Russia and other counties have archived some progress in creating the railguns and almost can construct the electromagnetic driver that is able to make the little objects (less than 1 kilo) have the orbital velocity and transfer goods to the orbit.

Under our project, the railguns should be placed near the Nuclear Power Station and the Hydroelectric power station, which allows to reduce the launch costs as it can use the excess of generated electricity (up to 30% at night).

2 Problem statement

We offer the technical and economic model under which the space power plants should be consisted of a number of standard modules that are launched through the railguns. This

* Corresponding author: loginovel@mail.ru

model defines the perspective integration of both space and ground electrical engineering as one super-system.

We also want to highlight the need for the package of monitoring methods based on new generation smart grid that provides receiving the information about the processes occurred in the energy systems [2]. This of monitoring methods assumes the possibility of analyzing the interactions between various subsystems of the integrated power super-system.

Limited transfer of energy from the orbit to the Earth provokes the lack of active power in many regions of Russia [3, 4].

The variety of manifestations of the effect of resonant stimulation (coherent resonance interaction) - from the spaced-coherent level to the integral system-structural reactions, on the one hand, and the loss of power objects from synchronism - the process of the output of the integrated power supersystem beyond the limits of the quasistable states.

On the other hand, with inevitability requires the construction of a single technical concept of self-balancing an integrated electric power supersystem.

The attempts to establish the correspondence between the stability of the vibrational levels of the "generation-transport-distribution-consumption" chain to achieve the local equilibrium point of the electric power super-system, determined by a combination of individual aperiodic or vibrational disturbances, did not lead to positive results, which, apparently, is due to insufficient consideration of the dependence on heterogeneity of the state parameters of the system objects, including a linear or nonlinear component that varies with time.

3 Solution

In the integrated electric power supersystem, when interacting with coherent oscillations generated by a substantially nonequilibrium active medium of separate segments of power grids and power energy objects with a large number of intelligent devices (with an inverse characteristic of the stability of the vibrational levels), self-balancing processes can occur, as a result which, in particular, will restore the functions of the damaged power soup a system that includes space and terrestrial systems.

The electric power supersystem, including space and terrestrial systems, is not only open, but also consists of active small objects, the structure of which is complex and insufficiently manageable.

Apparently, the concretization of the concept of an element of an integrated electric power supersystem essentially depends on the formulation of the problem and requires the consideration of collective interactions in the processes of energy transfer from different points of the geostationary orbits of space solar power stations to the ground.

The complex and non-linear dynamic behavior and incompleteness of knowledge about microstates in individual systems of the integrated electric power super-system make the application of coherent-resonant description unavoidable.

The possibility of using the effect of coherent resonance interaction of a quasi-integrated segment of the network (cloud, swarm) of standard modules as elements of space power plants as elementary entities is confirmed by the fact that the behavior of energy consumers and the need to take into account the influence of boundary conditions (contours supersystem) on the spectrum of admissible resonant states of a quasi-integrated segment in the chain of space and terrestrial systems "generation-transport-races limit-consumption ".

The presence of hierarchy in a multilevel structural organization of integrated space and terrestrial electric power industry as a single super system indicates the possibility of interconversion of dynamic synchronization areas (dissynchronization) of self-oscillations, as well as areas of technical activity in which stable and unstable cycles coexist.

Such transformations can be described taking into account the properties of the scale similarity between space and terrestrial energy systems.

Another aspect of the coherent-resonance interaction effect related to the phenomena of self-organization of integrated space and terrestrial electric power industry as a single super system is the adjustment of the spatiotemporal structure of dynamically changing equilibrium parameters to the generation-transport-distribution-consumption structure determined by a combination of individual aperiodic or vibrational disturbances, occurring due to the resonant properties of the object-technical structure.

Essential in this case is the fact that the resonant interaction formed in the structure of the "generation-transport-distribution-consumption" chain determined by a combination of individual aperiodic or vibrational disturbances is practically independent of the initial structure of the energy dynamics.

The attempts to establish the correspondence between the stability of the vibrational levels of the "generation-transportation-distribution-consumption" chain to find the local equilibrium point of a non-linear electric power supersystem including space and terrestrial systems did not lead to positive results, which, apparently, is due to insufficient consideration of the degree of "openness" integrated space and terrestrial electric power generation in the process of reactive power production / consumption both in power grids and in other segments integrated power supersystem and consumer space.

The changes in the resonant state of the quasi-integrated network segment (clouds, swarms) of standard modules as elements of space power plants, for example, their "clotting" or deployment, require calculating the parameters of reaching the local equilibrium point of a nonlinear super energy system, including space and terrestrial systems, and then finding the optimization maximum of this equilibrium state from the averaged parameters of the group of interacting terrestrial and cosmic systems, but lead to significant variations in their resonant activity.

The limited resource of the quasi-integrated segment of the network (clouds, swarms) of standard modules as elements of space power stations within the super system and the need to maintain objects sufficient for life in a given resonance state lead to the emergence during the evolution of several dynamic regions of synchronization (dissynchronization) of self-oscillations, in which a stable and unstable cycle coexist.

From the point of view of the super-system, this mechanism can be considered as a system of "synchronization (dissynchronization) of self-oscillations", which causes coherent-resonant nonequilibrium of the entire electric power supersystem, including space and terrestrial systems.

From a coherent-resonant point of view, such a disequilibrium can be considered as the cause of the coherent state in the chain of space and terrestrial "generation-transport-distribution-consumption" systems and interactions of various subsystems of the integrated electric power super-system and their ensembles, synchronization and suppression of some vibrational modes by others.

The main technical characteristic of the quasi-integrated segment in the chain of space and ground systems "generation-transportation-distribution-consumption", associated with its resonance state, can be considered self-oscillating moment.

The quasi-integrated segment in the chain of space and terrestrial systems "generation-transport-distribution-consumption" interacts with the nonlinear dynamical effect of energy quasi-harmonic oscillations, changing its stability to the vibrational state due to a change in the orientation and magnitude of its self-oscillating moment.

It should be noted that the self-oscillatory (multicyclic) moment of the quasi-integrated segment of the network (clouds, swarms) of standard modules as elements of cosmic power stations or its almost independent regions is generally nonzero, if only because of the presence of quasi-coherent technical fluctuations.

Since any technical system tends to minimize its potential energy, it can be assumed that the final effect of any coherent reaction must have a minimum self-oscillating moment of its own, if only because of the interaction of external and internal cycles of non-zero magnitude.

Therefore, in the process of coherent resonance reactions, objects have different self-oscillating moments and the super-system control operators can control their behavior and, at the end of the reaction, they mutually compensate as much as possible.

The self-oscillatory interaction of objects at resonance ensures the instability of the medium and the object, and also provides their interaction with coherent-resonant clusters of interacting factors of energy dynamics that is more effective in comparison with the final effect, which leads to stimulation of various resonance reactions.

This can be represented as follows: as a result of the self-oscillatory interaction of the quasi-integrated segment in the chain of space and terrestrial "generation-transport-distribution-consumption" systems with clusters of interacting factors, its oscillatory dynamics increases, which leads to an intensification of the process of "sorting out" the various resonant states and more rapid "finding" of a resonant state that is unstable for another object.

The disruption of the functioning of the super-system, apparently, primarily causes the accumulation of unnecessary effects of technical reactions.

Therefore, the growth of the self-oscillating moment in the objects leads to an increase in the system coherent resonance. In other words, the synchronization of the induced negative oscillations destabilizes the system or its segment and ensures the selectivity of the coherent-resonance interaction.

This simple technical consideration explains, in addition to the increased efficiency of the coherent-resonant interaction process for the unstable segments of the "generation-transport-distribution-consumption" chain, and some of the causes that lead to a difference in energy effects under conditions with a momentum "inside the system" and with a pulse "outward" system ".

Numerous data on the resonant stimulation of various effects also seem to be associated with a similar flattening of the spatial structure of the quasi- integrated segment of the network (cloud, swarm) of standard modules as elements of cosmic power stations, since the resonance center of many effects is also in the depth of a rather narrow segment.

The threshold value of the intensity of the energy dynamics, which still induces a coherent-resonant interaction, should be comparable with the degree of coherence of the nonlinear dynamic interaction of external and internal vibrational modes.

4 Conclusion

Elements of the electric power supersystem, including space and terrestrial systems, operate in the presence of multifactorial nonlinear dynamic interaction of external and internal cycles. Without specifying the control mechanisms of technical processes due to the object-technical coherent-resonant effects, it can be assumed that the multifactorial nonlinear-dynamic interaction determines the magnitude of the oscillations for such control systems.

The consideration of the complex analysis of data allows us to reveal the reasons of the massive dynamic disturbances in the super-system.

On this basis we can identify the self-balanced condition of the integrated power super-system and outline the functioning dynamics of the power subsystems.

Thus, clear understanding of the reasons and identifying the steps of the massive dynamic disturbances allows us to reveal the settings of the emergency control devices that will monitor the smart grid work if it faces some problems.

The theses were prepared with the financial support of the Russian Humanitarian Scientific Foundation (project No. 16-02-00463 a "Formation of organizational mechanisms for operating oil and gas resources on the basis of multi-agent modeling to protect Russia's economic interests from manipulating oil prices on world markets").

References

1. A. A. Makarov, T. A. Mitrova, F. V. Veselov, A. A. Galkina, V. A. Kulagin, Therm. Eng. **64(10)**, 703 (2017)
2. A. S. Bugaev, E. L. Loginov, A. N. Raikov, V. N. Saraev, S&T. Inf. Pr. **36(1)**, 68 (2009)
3. A. S Asochakov, Y. V. Shulgina, A. I. Soldatov, E. M. Shulgin, D. N. Ogorodnikov MEACS, (2015)
4. S.V.Chubov, A.I.Soldatov IOP Conf. Ser., MEACS **177(1)**, (2016)

Using an aerospace monitoring for sensing "lithosphere-atmosphere-ionosphere-magnetosphere" system in order to identify and outline the potential oil and gas fields

Alexander Ageev[1], *Svetlana* Bortalevich[2], *Evgeny* Loginov[2,*], *Alexander* Shkuta[3], and *Dmitry* Sorokin[2]

[1]RAS, Institute of Economic Strategies, 101000 Moscow, Russia
[2]RAS, Institute of Market Problems, 117418 Moscow, Russia
[3]Financial University under the Government of the Russian Federation, World Economics and International Business Department, 125993 Moscow, Russia

Abstract. The article focuses on consideration of complex aerospace monitoring of aggregated territories. It is made to assess the processes took place in "lithosphere-atmosphere-ionosphere-magnetosphere" system when exploration of oil and gas fields. The research is based on the remote aerospace sounding's method. As a result of this research we have an access to the complex analysis of huge range of data. We have an opportunity to calculate a "convolution operation" in the systems of complex aerospace monitoring in order to identify the signs that can help to indicate the territory having oil and gas deposits. To sum up, the research reveals the correlation between the huge range of anomalies of the nature and the presence of oil and gas deposits in the earth on the basis of remote aerospace monitoring sounding method.

1 Introduction

The technical capabilities of the distant aerospace sounding of the lithosphere and atmosphere have been expanded nowadays. At the same time, new approaches to the use of collected information on the ongoing physical processes have been identified by the researchers [1].

Because of the sounding of the lithosphere-atmosphere-ionosphere-magnetosphere system we now have an opportunity to identify and outline potential oil and gas deposits.

One of the most successful methods of exploration of oil fields is the organization of the distant aerospace monitoring.

Fixing various natural anomalies allows us to allocate areas of territories with potential oil and gas deposits. We have an opportunity to calculate a "convolution operation" in the systems of complex aerospace monitoring in order to identify the signs that can help to indicate the territory having oil and gas deposits.

* Corresponding author : loginovel@mail.ru

2 Problem statement

The estimation of the geological sphere can be realized based on the analysis of various kind of data received within the framework of a permanent monitoring of the situation in the lithosphere-atmosphere-ionosphere-magnetosphere system.

For this purpose it`s necessary to make tests on the prediction of electromagnetic disturbances which may took place in the lithosphere, atmosphere, ionosphere, magnetosphere and near-Earth space.

These tests results should increase the observability of the system and give the researchers an opportunity to calculate a "convolution operation" in the systems of complex aerospace monitoring in order to identify the signs that can help to indicate the territory having oil and gas deposits [2-4].

At the same time, it is necessary to reveal the correlation between the huge range of anomalies of the nature and the presence of oil and gas deposits in the earth based on remote aerospace monitoring sounding method [5, 6].

Various manifestations of seismic activity from individual tremors to earthquakes make the problem of predicting these very dangerous phenomena still extremely urgent.

In the last period of development of methods for remote aerospace sounding of the lithosphere and the atmosphere, the capabilities of the technical means of such sounding have expanded and, at the same time, new approaches to the use of collected information on the ongoing physical processes have emerged.

Based on the data from the sounding of the lithosphere-atmosphere-ionosphere-magnetosphere system, it is possible not only to predict earthquakes, but also to identify and outline potential oil and gas deposits.

3 Solution

One of the ways to improve methods for predicting seismic activity and exploration of oil and gas fields is the organization of remote aerospace monitoring sounding applied to the lithosphere-atmosphere-ionosphere-magnetosphere supersystem.

According to many authoritative experts, it is the physical processes that take place not only in the lithosphere, but in the supersystem, which unites the lithosphere, atmosphere and ionosphere, form the basis for interaction of oscillatory processes, the aggregate of which programs seismic activity.

Anomalies of different physical nature, recorded in the course of sounding during complex monitoring of territories, also make it possible to identify areas of areas with potential oil and gas deposits.

It is the complex nature of monitoring that, in contrast to individual methods of oil and gas exploration, allows us for greater observability of the studied characteristics of the territory, and, consequently, to expand the range of recorded parameters of physical manifestations that indicates the presence of oil and gas.

The dynamics of monitoring the metastable states of the geo-environment and their analogues within the framework of revealing the complex structure of the factors determining seismic activity allows us to identify a set of characteristics of coherent-resonant ensembles of vibrational modes as part of some transient activity of the lithosphere-atmosphere-ionosphere-magnetosphere supersystem.

Investigation of the properties of these associations (ensembles) of vibrational modes forms the possibility of predicting the exit of the geomedium beyond the limits of the quasi-stable states, depending on the inhomogeneity of the vibration parameters with their specific spatial configuration.

The processes of the geomedium exit beyond the limits of the quasistable states can be modeled in the formation of self-sustaining oscillations in it with a certain spatial configuration.

Moreover, chaotic peak manifestations of seismic activity can be represented as external manifestations of complex synchronization of quasiharmonic oscillations.

It is necessary to subdivide the transitional activity of the super-system "lithosphere-atmosphere-ionosphere-magnetosphere" in such a way that each of its aggregated segment from the upper mantle to the upper words of the atmosphere (taking into account the cosmic plaza) is represented as a sort of one macroobject.

This macroobject includes a number of evolving subsystems, each of which corresponds to the local state of the short-term relative equilibrium of the super-system, which can be characterized by the consolidating "convolution" of vibrational modes into ensemble-forming structures.

This state of the super system is strongly or loosely coupled with the coherent-resonant ensembles of the vibrational modes and is in substantial dependence on the processes of the formation in it of self-sustaining oscillations with a certain spatial configuration.

For the forecast of seismic activity, a package of models for monitoring natural and technogenic processes is required, incl. It is necessary to estimate the state of the geomedia and the probability of its escape beyond the limits of the quasistable states.

The estimation of the probability of critical dynamics of the geo-environment can be realized on the basis of the analysis of the dynamics of functional interrelationships of various geophysical factors on the basis of a complex analysis of heterogeneous data (including temperature and humidity profiles, electron concentrations, local parameters of ionospheric plasma, infrared radiation fluxes, etc.) monitoring refinement of the dynamically changing situation in the "lithosphere-atmosphere-ionosphere-magnetosphere" system.

Here, it is necessary to operate with working parameters of complex monitoring of aggregated territories. With the corresponding processing power, the obtained information on the processes occurring in the geosphere is processed in the context of the transient activity of the lithosphere-atmosphere-ionosphere-magnetosphere supersystem.

For this purpose, it is proposed to develop a set of tests for predicting electromagnetic disturbances that can be generated both in the lithosphere and in the atmosphere, the ionosphere, the magnetosphere and near-Earth space (cosmic plasma), on the basis of retrospective analysis and the current situation.

The test results should provide an opportunity to increase the observability.

As a result, a complex solution of the problems of the structural and functional organization of the possibility of predictive operation of the bifurcation parameters of the quasistationary states of the lithosphere-atmosphere-ionosphere-magnetosphere supersystem in the realization of the interrelationships between the metastable states of the geo-environment and geophysical processes determined by anomalies of various nature (in the lithosphere, in the atmosphere, in the ionosphere, magnetosphere, etc.), within the framework of the correlation between anomalies of different physical nature and seismic activity or correlations between anomalies of different physical nature and the presence of oil and gas deposits.

At the same time, a clear understanding of the causes and identification of phases of seismic activity makes it possible to isolate the characteristics of the fixed anomalies (in the lithosphere, in the atmosphere, ionosphere, magnetosphere, etc.) that identify the presence of oil and gas deposits.

4 Conclusion

As a result of this research we have an access to the complex analysis of huge range of data. There is a correlation between the geological sphere situation and the geophysical processes determined by the different natural anomalies (in the lithosphere, in the atmosphere, in the ionosphere, in the magnetosphere, etc.).

Another reviled correlation is between seismic activity, natural anomalies and the presence of oil and gas fields.

The theses were prepared with the financial support of the Russian Humanitarian Scientific Foundation (project No. 16-02-00463 a "Formation of organizational mechanisms for operating oil and gas resources on the basis of multi-agent modeling to protect Russia's economic interests from manipulating oil prices on world markets").

References

1. A. A. Makarov, T. A. Mitrova, F. V. Veselov, A. A. Galkina, V. A. Kulagin, Therm. Eng. **64(10)**, 703 (2017)
2. A. P. Grigorev, A. I. Soldatov, P. V. Sorokin *MTT*, 97 (2000)
3. D. N. Demyanovich, O. S. Vadutov, A. I. Soldatov MEACS, (2014)
4. A. S Asochakov, Y. V. Shulgina, A. I. Soldatov, E. M. Shulgin, D. N. Ogorodnikov MEACS, (2015)
5. A. S. Bugaev, E. L. Loginov, A. N. Raikov, V. N. Saraev, S&T. Inf. Pr. **36(1)**, 68 (2009)
6. E. L. Loginov, A. N. Raikov, Therm. Eng. **62(4)**, 233 (2015)

Applying the competence-based approach to management in the aerospace industry

Mariam Arpentieva[1,*], *Olga* Duvalina[1], *Svetlana* Braitseva[1], *Irina* Gorelova[2], and *Anna* Rozhnova[2]

[1]Tsiolkovskiy Kaluga State University, Psychology Department, 248023 Kaluga, Russia
[2]Russian Academy of national economy and public administration under the President of the Russian Federation, Volgograd Institute of management, 400131 Volgograd, Russia

Abstract. Problems of management in aerospace manufacturing are similar to those we observe in other sectors, the main of which is the flattening of strategic management. The main reason lies in the attitude towards human resource of the organization. In the aerospace industry employs 250 thousand people, who need individual approach. The individual approach can offer competence-based approach to management. The purpose of the study is proof of the benefits of the competency approach to human resource management in context strategic management of the aerospace organization. To achieve this goal it is possible to obtain the method of comparative analysis. The article compares two approaches to personnel management. The transition to competence-based human resource management means (a) a different understanding of the object of management; (b) involvement in all functions of human resource management «knowledge – skills – abilities» of the employee; (c) to change the approach to strategic management aerospace industry.

1 Introduction

The aerospace industry is one of the few that makes the Russian economy more competitive. A particularly important issue in the field of aerospace engineering is the question of the object of the management. The object of knowledge is a process, phenomenon, originating in scientific reflection in connection with their problematic situation. The number of such objects make management very difficult and un-successes. Un-differences make expert's and manager's positions ambiguous interpretation A priori, in the theory of strategic management, an organization was supposed to be an object, introducing strategic management into practice. Despite this, reality showed that, depending on the point of view on the essence of the organization, different management concepts was being form. This ambiguity led to the conclusion that «the heterogeneity of units of analysis» [6]. In the theory of strategic management is justified a two-level interpretation of the units of analysis, the object of strategic management. First, it is macro-object, it is organization. Secondly, it is a source of competitive advantage of the enterprise (micro-object). With the interpretation of competitive advantage, is not so simple. In the process of

* Corresponding author: mariam_rav@mail.ru

interpreting the notion of "competitive advantages", a much greater ambiguity is revealed. In fact, the evolution of the theory of strategic management is the evolution of the filling of the term "competitive advantages". The availability of physical resources has levelled out the possibility of obtaining a sustainable competitive advantage from the outside. What is inside the enterprise itself can be called a competitive advantage - the issue is much more complicated. Scientists are still arguing about him. But if we analyse the known approaches, the solution of this question is reduced in all the ways to the concept of "competence" [8].

2 Materials and Methods

The primary method is comparative analysis. Of particular interest is the benefits of implementing a competency approach to personnel management. Discusses various approaches to the desired concept are shown prospects for management based on the competence approach. The context of research is the strategic management in modern conditions. The strategic management theory sees the source of competitive advantage in dynamic abilities. Under the dynamic abilities of many researchers understand exactly competence. Researchers interpret the term «competence» in different ways [9].

3 Results

The transition to management based competency approach helps us: 1. diagnose the human resources of the organization (competence are the basis of the evaluation criteria resource); 2. to form a «competence passport» of employees and departments (this document helps to manage lifecycle of an employee); 3. to determine the place and role of each employee in implementation of the strategy of the organization; 4. identify priority areas of personnel policy; 5. to revise the emphasis in the incentive system in the enterprise; 6. determine the need and direction of training (retraining) of employees. This complex helps us in the implementation of strategic objectives.

4 Discussion

The etymology of the term «competence» comes from the Latin «competentia» as the membership rights. Used in the French language since the late XVIII century in the meaning of authority as confirmed right, authority. In the mid of the XIX century began to was use to denote a range of issues, in which anyone versed professional, with knowledge of the matter. The first part of the word (prefix com-) suggests that competence, as the authority must be confirm by something (a law, administrative act, power of attorney). The second part of the word (root base petere-) suggests, that competence as the authority may not appear effortless, it must be achieved, «summon». In the West, the revival of this notion an introduction it to the language theory (N. Chomsky, R. White) led to the transition to competence based education in the mid of the XIX century [7]. Further, the scientific community began work to identify the components of competence. A pioneer in this field was J. Raven. He identified 39 types of competences that equates with the «motivated abilities», a special place is given to self-education and self-development. In the modern management competence primarily defined as a «combination of knowledge, skills, motivational factors, personality traits, and situational intentions, which provides the effective solution of a task by a certain class in a certain organization at a certain place in a certain group» [10]. Thus, «competency» is inseparable from the personality, whereas «competence» - characteristics of the workplace or the working task. It should be remembered that (1) competence is a quality or property pertaining to professional activities. Competence can be consciously changed (2). That is why competence can be called dynamic, changing abilities. They can be transformed into specific needs of strategic management. In all situations and periods, the concept of "competence" is used by the Labor Code of the Russian Federation in the context of the notion of the powers of management bodies. Thus, the interpretation of the term in labor relations has not changed since the beginning of the XX century. The dictionary of foreign words issued at that time gives the definition of competence as the terms of reference of an institution or person. Modern Russian system of education is actively implementing a competence-based approach. In the education system under the purview understand the components of the work function in terms of «knowledge – abilities – skills». Professional dictionary is missing. The system of professional standards are not yet ready to answer the queries no education system, no system of labour relations. This state of things J. Baudrillard called simulation. [1] Simulation of a competence campaign in all spheres of life leads to tragic results (Figure 1).

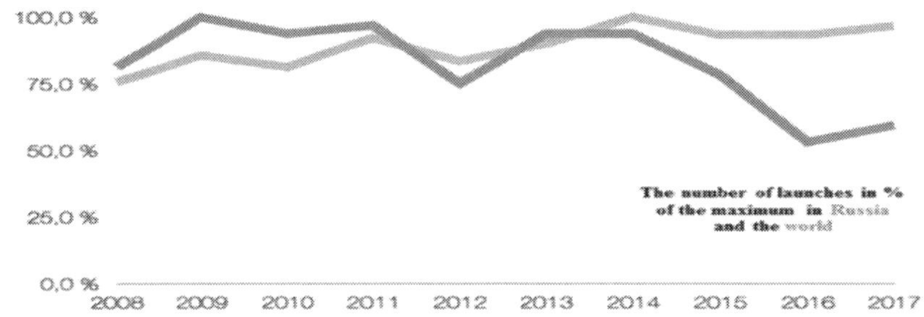

Fig. 1 Dynamics of space launches in Russia and in the world in 10 years [11].

The simulation results prove the use of competence approach in management (table 1). For implementing of competency in management, this term is required relevant interpretation of the object of management [2]. It involves the allocation of competences as micro-object in the management object. Macro-object stands the man himself. This understanding is reflected in the concept of human resource management, respectively, while all management functions associated with the development of the management object. We can distinguish a number of functions: 1. *Recruitment.* The search for a suitable candidate for a position is done by declaring the required competencies «knowledge – abilities – skills» which is included in the strategic management process and goal setting. This approach determines the versions of the methods of recruitment and given the reputational risk of the organization. 2. *Selection of personnel* is carried out as the process of identification of clearly defined requirements for the post and competence of the candidate. Hence, practitioners and theorists also have a choice of criteria and methods of selection. 3. *Regulation of labor relations* involves paperwork, which includes the future development of the employee (including in the reserve, career profile, reward system, involvement in voluntary insurance, etc.). The issue of compliance of labor legislation is not necessary. 4. *Certification of personnel* is carried out exclusively to identify the strategic potential of the organization. 5. *Staff training* is carry out based on the needs of the organization to achieve strategic objectives. The variability of the external environment suggests new requirements to human resources, determines the need of new competences. Hence, practitioners and theorists have ideas about the framework for the development of training programs, programs for individual development. From the idea of goals there is an understanding of the need to spend money on this aspect of management. 6. *Motivation and stimulation* of the personnel is form based on strategic needs and opportunities of the organization. The creation was motivate by the need for competence is stimulated by the degree of its involvement in the strategic process. 7. *Downsizing of staff* is in accordance with the strategic needs of the resource of the enterprise.

Table 1. Results of the compensation approach: simulation and implementation.

Simulation	Competence-based approach	Implementation
The enforcement of the decision. The principle of operation in the organization: «There are two opinions regarding the solution to the problem. One solution belongs to guide other a priori wrong»	Tool to reach consensus/consent	In the strategic management means understanding the alignment of interests
The prevailing mechanism of self-affirmation, which eliminates the possibility of recognizing user errors	Objective function: strategic management the essence of setting a goal and achieving it	Development of goals, mission , vision control system, staged its achievement, taking into account permanent error analysis
Management in the academic sense is missing. Replace his powers, disappears the subject – object relationship Failures justify human factor and national model.	Management tools	Rational mechanisms and procedures which are implemented subject – object relationship
Subjective and irrational, imposed, external environment conserved its manifestations are spontaneous, unpredictable, depend on the personality	View of the external environment	Objectively, comes from rational objectives, opportunities, resources, and environment limitations
The imposition of representations of the environment requires an explanation of any changes (excuses, guessing)	Attitudes towards changes in the external environment	Flexibility in understanding the «life cycle» environment, causes of changes
Not required	Delegation of functions	Delegation is one of the important functions and conditions for achieving strategic goals
Actual sentence: "In the struggle all means are good".	The choice of management tools	Different tools
Attitude as a means of obtaining benefits. In the development of the resource issue is not	Related to human resource	Development people in accordance with the analysis of existing competencies
Not required	Attitude to feedback	Feedback is needed as the alarm system on the destructive aspects in the achievement of strategic goals
Not required	Requirements for the subject	Requires self-organization Manager

5 Conclusions

The formation of a competence approach in the organization is similar to the search for a philosopher's stone. Scientific and methodical literature devoted to the subject under consideration is full of numerous methods. These methods are mainly declarative (simulative from J. Baudrillard), a formalized character. At their core, they are divorced from the object of managerial influence: from the employees of the organization. They were burden with complex terminology (competencies, grading, etc.). In the United States in competence-based pay great attention to the analysis of the description tasks (competence exists as a workplace). Description of tasks can was call proto-elements of competencee. Especially hard, particularly painstaking work was conduct by scientists in the direction of a unified, unambiguous interpretation of terms and concepts in the methods used. The seriousness of the plans confirms the fact that even reward systems were patent in the US. These systems were recognize as unique developments that allow the effective management of the company. In Russia at least, such a systems was impose to them. These systems were recognize as a commercial secret. However, they did not always meet the requirements of modern management. Examples of qualitative methods in the context of a competence approach can be those based on a factorial approach. Especially popular among them among management managers and researchers was the method of E.N. Hey (grading) [5] In the Soviet Union, scientists developed a similar project. They singled out the basic principles for assessing the quality and complexity of the work of experts (for determining the categories of qualifications and for differentiating salaries). Developed these recommendations of the Institute of Labor (on behalf of the USSR State Committee for Labor and Social Affairs). [4]. The foreign design of management functions and the results of Soviet researchers are very similar. Unfortunately, many achievements of the world and domestic science are lost in endless reforms. Theories and technologies that could help people form and develop interpersonal relationships are lost. These are theories and technologies of command formation and development of the collective. A look at the described problems through the concept of binary concept "simulation - dissimulation", "monosubjective - intersubjective", "focused - defocused", helps us to understand the true mechanisms of what is happening in modern management of aerospace objects [12].

References

1. J. Baudrillard. *Simulyakri i simulyatsiya* [In Russian] (Tula: Tulskiy poligrafist, 2013)
2. V. Demyanenko. *Osobennosti upravleniya izmeneniyami v Rossii* [In Russian] (Moscow: Publishing solutions, 2016)
3. I.V. Gorelova *Kompetentnostnyy podkhod k upravleniyu v Rossii: modelirovaniye ili dissimilyatsiya*. [In Russian] (Belgorod: BSI, 2017)
4. *Rekomendatsii po otsenke slozhnosti i kachestva raboty spetsialistov* [In Russian] (Moscow: Economics, 1989)
5. R. Henderson. *Compensation management in a knowledge-based world* (Upper Saddle River: Pearson Prentice Hall, 2006)
6. V.S. Katkalo. *Evolyutsiya teorii strategicheskogo upravleniya* [In Russian]. Author's abstract diss. of the doctor of economic sciences. – St.Petersburg (2007)
7. Y. S. Kostrova. Genezis ponyatiy «kompetenstiya» i «kompetentnost'» [In Russian]. *Young scientist* **12(2),** 102 (2011)
8. A. F. Moskovzev (ed.). *Upravleniye razvitiyem strategicheskogo potentsiala regiona* [In Russian] (Volgograd: VSTU, 2015)
9. G. Mintzberg, B. Ahlstrand, D. Lampel. *Strategy safari: A guided tour through the wilds of strategic management* (London: The Free Press, 1998)
10. A. Ovchinnikov. O klassifikatsii kompetentsii [In Russian]. *Organizational psychology* **4(4),** 145 (2014)
11. V. A. Zhartun. Kosmicheskaya katastrofa [In Russian]. *Sulakshin Center*, **14 January,** URL http://rusrand.ru/analytics/kosmicheskie-avarii (accessed: 18.01.2018) (2018)
12. V. A. Vittikh. Evolution of Ideas on Management Processes in the Society: From Cybernetics to Evergetics. *Group Decision and Negotiation* **24,** 825 (2015)

Didactic communication in the training of specialists in aerospace engineering

Mariam Arpentieva*, *Peter* Menshikov, and *Svetlana* Braitseva

Tsiolkovskiy Kaluga State University, Psychology Department, 248023 Kaluga, Russia

Abstract. The article is devoted to the study of the problems of didactic communication in the training of engineering personnel for the aerospace industry and to the study of the problems of the communication of subjects concerning the training and education of highly qualified engineering personnel for the aerospace industry. In the training of engineering personnel for the aerospace industry the integrated model of didactic communication involves the identification and description of its various components, typical modes of interaction (modes) that reflect different aspects of the person's understanding of the world around him and himself in the process of different types of education and upbringing. Didactic communication in the process of training engineering personnel for the aerospace industry is a multi-level, multi-stage and multi-component phenomenon. The modes, possibilities and limitations of this communication are related to the level and direction of personal, interpersonal and professional development of interaction subjects. The productivity of preparing engineering personnel for the aerospace industry is related to the choice of a model of didactic communication, which is addressed in different ways to the development of cognitive, value-semantic and meta-cognitive structures that form one or another type of education and upbringing.

1 Introduction

Didactic communication is the communication of subjects in the process of training and educating engineering personnel for the aerospace industry. It is help to achieve and develop understanding (comprehension) of learning phenomena of the surrounding world and themselves. In the training of personnel for the aerospace industry, didactic communication serves to the solution of educational and cognitive tasks. Didactic communication is also a means of communication, exchange of ideas, experiences between the learner and students in the course of joint activities. It helps to overcome difficulties and barriers in the process of educational and cognitive activity. It helps to activate the establishment of contact with students, provides for their more active involvement in the process of cognition, partnership in the extraction, transfer, processing of educational information [1-7]. Autodidactic communication includes communication about the self-learning of engineering personnel for the aerospace industry, its components and processes.

* Corresponding author: mariam_rav@mail.ru

2 Materials and Methods

The primary method is theoretical integrative analysis. Didactic communication in the process of training engineering personnel for the aerospace industry is a multi-level, multi-stage and multi-component phenomenon. Its modes, possibilities and limitations are related to the level and direction of personal, interpersonal and professional development of interaction subjects. These characteristics of the development of future specialists as subjects of didactic communication in the educational and professional activities manifest in several field of engineering in the aerospace industry. They manifest in the peculiarities of their understanding of themselves and the world around them, the situations (tasks) of these activities, the development of reflection and information technologies and structures involved in solving problems of professional and educational activities.

3 Results

The main features of didactic communication in the process of training engineering personnel for the aerospace industry are related to its modes (orientation): teaching, learning, self-learning (self-teaching), mutual learning (mutual teaching). Didactic communication in its various modes has different capabilities and limitations in the training and development of interaction subjects, in many respects regardless of the subject area or specialization to which the training specialist belongs.

1. Didactic communication in each of the modes is aimed at the formation and development of competence and professionalism, the formation and development of harmonious human relations with oneself and the world. Didactic communication is the key to the success (efficiency and effectiveness) of the activities of teachers and students in the process of training of engineering personnel for the aerospace industry. In each mode, the processes of (re)translation and transformation of knowledge and skills, values and meta-knowledge of different subject-professional areas (as components of professional activity) are expressed in varying degrees and specific forms. These processes are realized by means of both similar, general and specific psycho-technologies, which form a certain level and type of the ability to learn and the ability to teach, the ability to educate and the ability to be brought up.

2. Didactic communication in the process of training engineers for the aerospace industry in each of its modes is associated with the primary implementation of certain types of training (for example, "superficial", reproductive, and "deep", creative). The development of didactic communication was determine by the development of values and metacognitive abilities of the subjects of communication. This manifest, among other things, in changes in the understanding of the self and the world around us. This manifested in transformations of the way (type or level) of understanding oneself, the interlocutor, and the situations of educational and teaching activities, world outlook in general. This manifested in the general growth of professional and personal competence, maturity as readiness for productive and harmonious functioning in the sphere of professional and non-professional relations. In the preparation of engineering personnel for the aerospace industry, all the marked modes are represent, although in varying degrees. Self-learning and mutual learning (mutual teaching) is a part of the educational process and part of the processes of labor efficiency. Self-learning and mutual learning are increasingly significant.

4 Discussion

In the process of training engineering personnel for the aerospace industry, the development of didactic communication is part of the personal and interpersonal development of communication subjects. This development is connected with the development of values and metacognitive abilities of the subjects of communication. It is reflected in the deepening and expanding understanding of the person and the world around him. It is reflected in changing the type and level of understanding of himself, the interlocutor, the situations of educational and educational activities, the process of education and peace in the world. Autodidactic communication is formed. The "understanding ability" of the individual in relations with himself and with the world develops was strengthen and purse. The dialogue intentions was strengthen and purse too. The interlocutors develop an aspiration for dialogic communication. There is an intensification and realization of dialogical intentions. There is a growing desire for dialogic relations. Didactic communication is presented as an mutual understanding process. It is aimed at the dialogical understanding with a "Significant Other" (such as the teacher or specialist) in the process of training engineering personnel for the aerospace industry. Such communication develops the framework of educational and professional interaction, becoming a dialogue of life-worlds. In the process of training engineering personnel for the aerospace industry a natural change in the modes of didactic communication occurs. This change include the conscious and unconscious accumulation of knowledge and skills, psychotechnologies and values, personal maturation and the professional development of man. In parallel, the modes of instruction are changed. From the first type (passive-reproductive "learning") and the second type (actively-contextual "learning"), the person passes to the third type (self-learning) and the fourth type (dialogic, mutual learning/ mutual teaching).

Each of the types of learning is thus associated with this or that type of didactic communication:

1. Teaching didactic communication assumes that the teaching of young men and women - future engineers in the aerospace industry – is connected with the movement from the learner to the learner. This is an insignificantly realized appropriation of knowledge and skills broadcast by the instructor (teacher). Knowledge and skills in this case are understood with greater or less completeness and depth, not being the subject of care and reflection of subjects of communication. The meta-cognitive aspects of teaching and studying are inactive. Typically, rather unrecognizable appropriation of psycho-technologies, goals and values of activities associated with the appropriated subject-specific (not yet professional-specific) knowledge and skills; typically unrecognized appropriation of the components of learning activity (the ability to learn). Knowledge and skills of learning activity, values and psycho-technologies, characterized by a general passivity of learning. Within the framework of this mode, to a greater or lesser extent, the surrounding world, its various components, was comprehend. This is a pre-professional stage of formation, in which the ability to learn is partly formed, changing behavior in the process of learning, and "becoming a student" occurs. Understanding is objective, reproductive-ascertaining nature (ascertaining explanations or, in terms of humanistic psychology, "objectively knowing") [4, 6, 8].

2. Teaching didactic communication assumes that the training of youths and adults, nonprofessionals and professionals is addressed not only to the appropriation of knowledge and skills, but also to change the behavior of the trainee on their basis, the implementation of (quasi-) professional activity. At this level, there is a more or less conscious transfer of knowledge and skills. This knowledge and skills act as a support for the organization and implementation (change) of activities. There is realized (re)translation of values, psycho-technologies and meta-knowledge quasi-professional activities. This mode as a whole is

relatively unrecognized. Active development of the components of educational, educational and professional activity is typical. An understanding of oneself is realized through an understanding of the educational and professional and professional activities. The requirements of these activities allow us to assess ourselves in its context, determining the measure of compliance with its requirements, its "objective success". This is the initial level of the emergence of professionalism, in which the individual faces the task of changing in accordance with the requirements of activity, developing the ability to learn, that is changing activity and oneself in the learning process. The realizing type of understanding is associated with the predominance of subjective, in many respects subjective, personified explanations, here an empathic-experiential modus (in humanistic psychology, denoted by the term "interpersonal knowledge" - "interpersonal knowing").

3. Didactic communication in the process of self-education (autodidactic communication) involves a qualitative transformation of educational, educational, professional and professional activities in the process of training engineering personnel for the aerospace industry. In this modes come to the fore the more or less conscious, detailed and profound comprehension and transformation of the components of educational, educational, professional and professional activities. This is training to training. There is, therefore, a transformation of the activity of a person educating himself on the basis of skills and habits acquired and transformed by him in the process of development. There is a transformation of his values and goals, psycho-technologies and meta-knowledge of one or another educational, professional and past-professional activity. In the context of this mode, realized, selective mastering of values, psycho-technologies and meta-knowledge (quasi-)professional activities. These processes are realized in different ways and ensuring the implementation and qualitative development of activities in general (the transformation of the activity and life activity of the subject). The change in the data of the organization's "supports", the implementation and transformation of activities is associated with the selective search and development of educational and professional knowledge and skills. The specialists are being realized the relatively conscious, detailed (multi-component) and profound reintegration of the knowing and skills. A person begins to understand himself as an inseparable and changing integrity, each component and process significance for himself and the world. Achieving self-understanding is a condition for deep and extensive communication with the world. The ability to learn reaches the status of the ability to change, develop, changing the parameters and technologies of internal and external activities. This is one of the important steps in the way of becoming a person by himself. "Becoming a person", "becoming a professional" mean reaching the state of the self-efficacious and able to teach not only himself but others. These personal levels mean reaching the state to be perfected in the process of self-learning and learning of others (self-efficacy) [4-6]. At the professional level a person chooses the direction of his development, proceeds to a creative rethinking of his professional activity. A type of understanding characteristic of a given level can be designated as an interpretation, less often as a dialog of the experience of experiences (field of experience, emergent, arising in an event and transpersonal cognition) [1, 6] .

4. Dialogical didactic communication in the process of training of engineering personnel for the aerospace industry is characterized by mutual, in different measure, conscious learning by people of each other. This suggests the ability to change one's own and those of others (life-activity) in the process of exchanging the values of this life exchange of values, psycho-technologies and meta-knowledge, as well as knowledge and skills of interaction. This knowledge and skills were experience, understood by the person as changing. They and their meanings change in relation to activity, and in relation to the reality (world). Learning to learn, life-related changes in life is the leading aspect of the development of the life of the subjects in general. Mutual learning changes the learning of each other as

individuals, changes their lives in the process of developing awareness (change) of oneself and the world. A person begins to understand himself and the world in an indissoluble unity: as an inseparable and changing integrity, every component and process of internal and external life has equal importance for man and the world. Such a state of "flow" is described as a state of "self-efficacy" (in the behavioral tradition) and as a state of "full functioning" and self-actualization, "becoming a self, partner, professional," "being" or "meeting" (in existential- humanistic psychology) [1, 4, 7].

5 Conclusions

A separate kind of didactic communication is mentoring and supervising [5]. Here we are talking about didactic communication in the process of training and retraining of engineering personnel for the aerospace industry in the framework of their own professional activities. In general, such communication, as a rule, can be fragmented and applied to the context of previous modes: in models of learning, learning, self-education, where the weight of metacognitive aspects of teaching and upbringing is most significant.

The transformation of the didactic system includes, thus, the transformation in the process of training engineering personnel for the aerospace industry:

1) the didactic system and its components (changing the type of (re)presentation of knowledge and skills, psycho-technologies and meta-knowledge, changing the values of the didactic system);

2) the ability to learn (transition to a new level of learning, the transformation of values and the psycho-technologies of instruction);

3) professional activity (change of values, psycho-technologies of activity, search and acquisition of knowledge and skills);

4) the personality and its relations with the world, the didactic system as a whole (values, psycho-technologies of learning activity, (meta) knowledge).

References

1. M. R. Arpentieva, P. V. Menshikov, *Didactikticheskaya kommunikastiya: umenie uchit'sya I umenie uchit'* [In Russian] (Kaluga: KSU, 2017)

2. P. V. Menshikov, *Phenomenon didactikticheskoi kommunikastii* [In Russian] (Kaluga: KSU, 2017)

3. P. V. Menshikov, M. R. Arpentieva, Novie obrazovatel'nie praktiki v kontekste kommunikativnogo podhoda [In Russian], *Vocational education in the modern world* **3**, 1179 (2017) DOI: 10.15372 / PEMW20170306

4. G. Bateson, *The Ecology of Reason* [In Russian], (Moscow: Sense, 2000)

5. C. R. Rogers, *Freedom to learn* (Columbus: Merrill, 1969)

6. C. R. Rogers, *Becoming partners* (N. Y.: Delacorte Press, 1972)

7. C. R. Rogers, H. C. Lyon, R. Tausch, *On Becoming an Effective Teacher* (London: Routledge, 2012)

8. H.-J. Bak, & D.H. Kim, Research in Higher Education **56,** 843 (2015)

Design issues of the piezo motor for the spacecraft reflector control system

Anton Azin[1,*], *Sergey* Rikkonen[1], *Sergey* Ponomarev[1], and *Stanislav* Kuznetsov[2]

[1]National Research Tomsk State University, Research Institute of Applied Mathematics and Mechanics, 634050 Tomsk, Russia
[2] Academician M.F. Reshetnev Information Satellite Systems,662972 Zheleznogorsk, Krasnoyarsk region, Russia

Abstract. Creation of large-size reflectors for spacecrafts is a topical issue for the space industry. The accuracy of the reflecting surface form and the structure weight are the main criteria for the reflector design. The accuracy of the reflecting surface form during a long-term operation is provided by adjustment when using piezoelectric motors in the reflector design. These motors have small weight-size parameters and can reach great torque values. The piezo motor is a distributed mechanical-acoustic oscillation system. Mechanical-acoustic oscillations are generated in the piezo motor by a PZT-stack and transmitted to an oscillator element, and then from the oscillator element to a load action element. At high frequencies, when dimensions of the oscillator are proportionate to the wavelength, the energy is transmitted by means of acoustic waves. In this case, mechanical waves practically are not involved in the energy transmission process. This thesis shows a method for selecting the material of a mechanical-acoustic oscillation system according to the efficiency of the acoustic energy transmission via a piezoelectric layered structure.

1 Introduction

Reduction of weight-size parameters of the spacecraft (SC) systems is the key issue in the space industry. One of the solutions is substitution of electromechanical drives of different SC devices for a variety of piezo motors, while weight-size parameters of the corresponding devices are reduced by several times. PZT-stack based piezo motors are the most reliable and powerful ones [1, 2].

The piezo motor is a distributed mechanical-acoustic oscillation system. Mechanical-acoustic oscillations are generated in the piezo motor by a PZT-stack and transmitted to an oscillator element, and then from the oscillator element to a load action element. At high frequencies, when dimensions of the oscillator are proportionate to the wavelength, the energy is transmitted owing to acoustic waves. In this case, mechanical waves practically are not involved in the energy transmission process.

* Corresponding author: antonazin@niipmm.tsu.ru

2 Determine of the efficiency of the plane wave acoustic radiation in the multilayer structure

Acoustic waves are a phenomenon characteristic for solid, liquid and gaseous media. The acoustic wave transmission (in general, oscillations) occurs systematically in full accordance with vibration of elastic bodies described by the oscillation equation. The acoustic wave transmission is described by the oscillation equation. The energy transmission efficiency is evaluated by the physical concept of the acoustic impedance [1-4].

Within the medium that does not contain external acoustic sources, the equation for a plane wave distributed in X direction will be:

$$(d^2/(dx^2)+k^2)\Phi = 0, \quad (1)$$

where: $k = \omega\cdot\sqrt{(\rho/\varkappa)} = \omega/c = 2\pi/\lambda$ – wave number, $\varkappa$ – Young's modulus, λ – wave length, ω – angular frequency of rotation, c – acoustic velocity, Φ – efficiency of the source related to a unit volume by the velocity [5].

To determine the energy transmission efficiency of a travelling plane wave, it is necessary to calculate the acoustic impedance (2).

$$Z = F/\dot{X} = \rho\cdot c = \sqrt{\rho\cdot\varkappa}, \quad (2)$$

where: ρ and $\varkappa$ – density and bulk elasticity coefficient (Young's modulus) of the layer, $\dot{X}$ – velocity potential in the layer, F – force of influence of one medium on another. A plane wave is an ideal mechanism of the energy transmission. Let us consider the task of the plane waves, distribution in a layered medium formed by parallel to each other layers of the substance (Fig. 1).

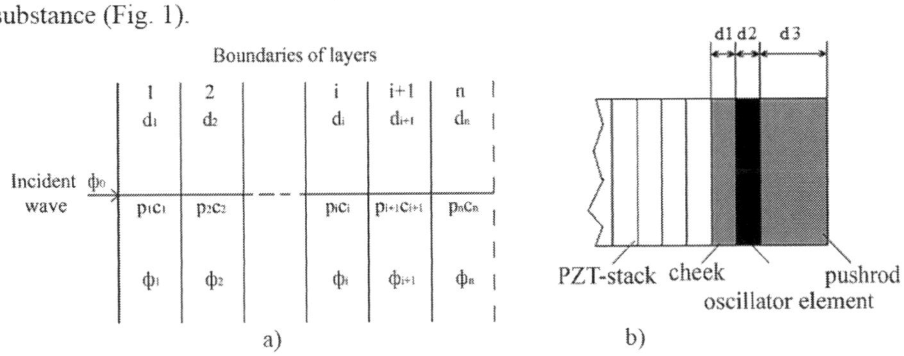

Fig. 1. Distribution of a plane wave in the multilayer structure: a) distribution of a plane wave in the n-layer medium; b) oscillation system of the piezo motor in terms of the multilayer structure.

The task solution for a three-layer structure will define the efficiency of the acoustic power η transmission. Let us calculate a ratio of the output signal W3 (the plane wave power in the third layer) to the input signal W0 (the plane wave power in the first layer) [5].

$$\eta = W3/W0 = (4R_1\cdot R_2^2\cdot R_3)/((R_1+R_3)^2\cdot R_2^2\cdot \cos^2\theta_2+(R_2^2+R_1\cdot R_2)^2\cdot \sin^2\theta_2). \quad (3)$$

Here $R_i = \rho_i\cdot c_i$, $\theta_2 = k_2\cdot d_2$, d_2 – thickness of intermediate second layer.

The piezo motor design and materials are considered in terms of the multilayer structure under conditions of acoustic radiation (Fig. 1) [3, 4]. Materials with the low acoustic impedance are plexiglass and bakelite. Aluminium and titanium have the lowest acoustic impedance among metals.

The material of cheeks, the oscillator element and the pushrod may vary, but at that time, the process of acoustic radiation of the plane wave will occur with a different efficiency.

The equation (3) is used for calculation of the efficiency of the plane wave acoustic radiation for materials with different wave impedances.

In the first case under consideration, aluminium is applied as a material for the cheek, oscillator element and the pushrod. The wave impedance of aluminium is $R_{aluminium} = \rho_i{\cdot}c_i{=}1.377{\cdot}10^7$ [kg/sec·m²]. The calculation shows that the piezo motor efficiency for identical wave impedances is equal to one, and the oscillator element length does not affect the radiation efficiency.

In the second case under consideration, aluminium is taken as a material for the cheek and the oscillator element, and the pushrod is made of plexiglass. The wave impedance of plexiglass is $R_{plexiglass} = \rho_i{\cdot}c_i = 3.0{\cdot}10^6$ [kg/sec·m²]. In this combination of materials, the system efficiency was reduced to 0.6 (60 %) in the low frequency area and became frequency dependent. With increase in the thickness of the oscillator element layer, the system efficiency gets low and reaches 0.2 (20 %) at frequencies above 100 kHz (Fig. 2).

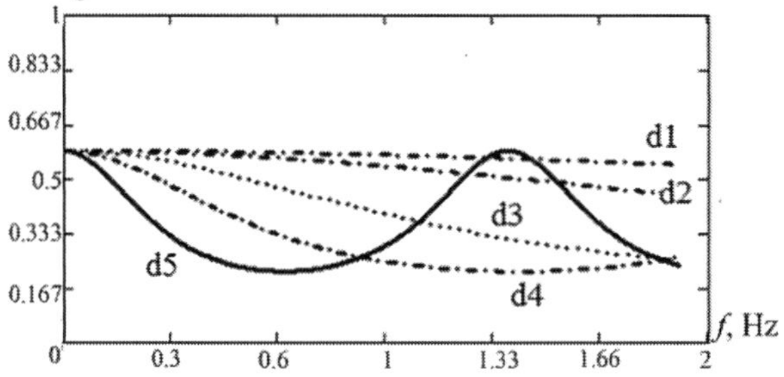

Fig. 2. Frequency response curve of the acoustic system efficiency. A material of the cheek and the oscillator element is aluminium; a material of the pushrod is plexiglass. The thickness of the oscillator element layer is: d_1 =0.92 mm, d_2 =1.8 mm, d_3 = 4.6 mm, d_4 = 9 mm, d_5 = 18 mm.

In the third case under consideration, plexiglass is taken as a material for the cheek and the oscillator element, and the pushrod is made of bakelite. The wave impedance of bakelite is $R_{bakelite} = \rho_i{\cdot}c_i = 2.06{\cdot}10^6$[kg/sec·m²]. The wave impedance of plexiglass and bakelite are similar and the system efficiency with a small thickness of the oscillator element layer is about to one. Increasing the thickness of the oscillator element layer will reduce the system efficiency up to 0.7, and for the thicknesses above 9 mm the efficiency is variable depending on the excitation frequency (Fig. 3).

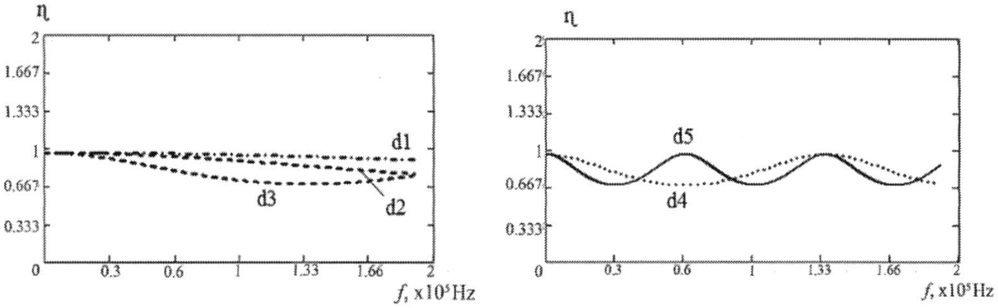

Fig. 3. Frequency response curves of the acoustic system efficiency. A material of the cheek and the oscillator element is plexiglass; a material of the pushrod is bakelite, and with the different thickness of the oscillator element layer is: d_1 =0.92 mm, d_2 =1.8 mm, d_3 = 4.6 mm, d_4 = 9 mm, d_5 = 18 mm.

In the fourth case under consideration, Bakelite is taken as a material for the cheek; the oscillator element and the pushrod are made of plexiglass. According to the calculation done, it is determined that the system efficiency is 0.96 and does not change for different thicknesses of the oscillator element layer.

3 Conclusion

The developed calculation procedure allows for qualitative selection of materials for the pushrod elements design with the maximum efficiency of the acoustic energy transmission. To obtain the maximum efficiency, it is necessary that the wave impedances will be similar or close by their value; in this case, the efficiency is about to one. When the thickness of the oscillator element layer is increased, the system efficiency is reduced up to 0.7; and for the thickness above 9 mm, the energy transmission efficiency is variable depending on the excitation frequency. The oscillator element thickness should be minimum and determined by the pushrod design peculiarities.

This work was financially supported by the Ministry of Education and Science of Russia; unique identifier RFMEFI57817X0257.

References

1. S. Park, S. He, Ultrasonics **52**, 880 (2012)
2. Z. Wang, Aerospace Science and Technology, 160 (2013)
3. A.V. Azin, E.P. Bogdanov, S.V. Rikkonen, S.V. Ponomarev, A.M. Khramtsov, IOP Conf. Ser.: Mater. Sci. Eng **177** 012003 (2017)
4. A.V. Azin, S.V. Rikkonen, S.V. Ponomarev, A.M. Khramtsov, *Proceedings of the Scientific-Practical Conference "Research and Development - 2016"*, 247 (2017)
5. T. Hayasaka *Electroacoustics* (Moscow: publishing house WORLD, 1982)

Mathematical model of a fluxgate magnetometer

Pavel F. Baranov*, *Vitalia E.* Baranova, and *Tamara G.* Nesterenko

National Research Tomsk Polytechnic University, 634050 Tomsk, Russia

Abstract. In paper analytical equations for calculate the electromotive force in the measuring coil of the fluxgate magnetometer independent of the drive signal frequency content are presented. Also, the equations for estimation of the fluxgate sensitivity at any harmonic and for study fluxgates operation with a glance to the waveform and the polynomial approximation of the mean magnetization curve of the core are provided.

1 Introduction

The principle of a fluxgate operation is based on the change in the ferromagnet state due to the impact of two magnetic fields with different frequencies [1-3]. A simplest fluxgate configuration represents a ferromagnetic rod which allocates an AC driving coil and a measuring coil [4]. Due to the variable magnetic field created by the driving coil current, a magnetic core goes through the magnetic reversal along a symmetrical curve.

A change in the magnetic flux caused by the magnetic core reversal along a symmetrical curve, induces the electromotive force in a measuring coil which varies in accordance with the simple harmonic motion. Under the effect of the constant or slightly varying magnetic field affecting the magnetic core, the magnetic reversal curve changes its size and shape and becomes asymmetrical. This, in turn, modifies the value and harmonic composition of the electromotive force in the measuring coil. In particular, there appear even electromotive force harmonics the amount of which is proportional to the magnetic intensity. These even harmonics are not observed on a symmetrical curve of the magnetic reversal.

The magnetic intensity is a vector variable, and core magnetization depends not only on its orientation but also a correlation between its longitudinal and transverse sizes. Therefore, the proposed fluxgate is characterized by a direction diagram and thus can be used for measurement of the vector angles and components of the flux density.

In order to create a high-sensitive fluxgate magnetometer [5], it is advisable to analytically describe its operation and create a mathematical model.

2 Operating principle of a single-rod fluxgate

A simplest single-rod fluxgate, the schematic representation of which is given in figure 1, measures a constant magnetic field.

* Corresponding author: bpf@tpu.ru

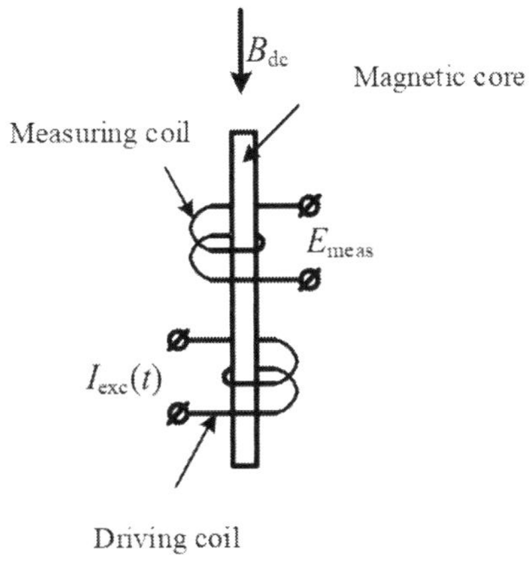

Fig. 1. Schematic of single-rod fluxgate.

Under the alternating exciting field $H_{exc}(t)$ which envelopes the magnetic core and the constant field H_{dc} directed along the magnetic core, the magnetic intensity superposition of these fields occurs in the core bulk:

$$H_{\Sigma}(t) = H_{dc} + H_{exc}(t). \tag{1}$$

The flux density in the core can be then obtained from

$$B(t) = f\left(H_{\Sigma}(t)\right) = f\left(H_{dc} + H_{exc}(t)\right). \tag{2}$$

The equation for the electromotive force in the measuring coil is as follows:

$$E_{mes}(t) = -sw_2 \frac{d}{dt} B(t), \tag{3}$$

where s is the area of the core transverse section, mm^2; w_2 is the number of turns in the measuring coil.

The flux density $B(H)$ of the uniformly magnetized magnetic core in the first approximation (without regard for anisotropic and hysteresis phenomena) is usually calculated using the core magnetization function.

When the core undergoes a periodic magnetic reversal, a good agreement between theoretical calculations and experimental data is achieved for the fluxgate operation. In this case we use the mean magnetization curve which is computed by the following equation:

$$B(H) = \left(B\uparrow(H) + B\downarrow(H)\right)/2. \tag{4}$$

Arrows in this equation denote the flux density at ascending and descending branches of hysteresis loops. The use of the mean magnetization curve rather than the normal, is explained by the fact that the optimum operating mode for the fluxgate is the core magnetic reversal along the hysteresis loop whose ascending and descending branches slightly differ from that of the mean magnetization curve. The analysis of the fluxgate operation utilizes the polynomial approximation of the mean magnetization curve.

As the mean magnetization curve is an uneven function, it can be approximated by the uneven degree polynomial:

$$B = \sum_{i=1}^{j} a_{2i-1} H^{2i-1}, \tag{5}$$

where a_i is the i-th approximation coefficient.

For the analysis of the fluxgate operation, the dependence between the magnetic flux density B and intensity H is commonly approximated by a third-order polynomial:

$$B = a_1 H + a_3 H^3. \tag{6}$$

Using (2) and (6), we can determine the magnetic induction in the core:

$$B(t) = a_1 H_{dc} + a_1 H_{exc}(t) + a_3 H_{dc}^3 + 3a_3 H_{dc}^2 H_{exc}(t) + 3a_3 H_{dc} H_{exc}^2(t) + a_3 H_{exc}^3(t). \tag{7}$$

When the sinusoidal alternating current excites the fluxgate, i.e. $I_{exc}(t) = I_{exc \cdot m}\sin(\omega t)$, the equation for the magnetic intensity can be written as

$$H_{exc}(t) = H_{exc.m}\sin(\omega t) = \frac{w_1}{l} I_{exc.m}\sin(\omega t), \tag{8}$$

where $I_{exc \cdot m}$ is the amplitude of excitation current, A; w_1 is the number of turns in a driving coil, m; ω is the frequency of excitation current, rad/s.

Substituting (7) into (3) in view of (8), we obtain the electromotive force in the measuring coil of the single-rod fluxgate:

$$E_{mes}(t) = \left[\left(a_1 + 3a_3 H_{dc}^2 + \frac{3a_3 H_{exc.m}^2}{4}\right)\omega H_{exc.m}\cos(\omega t) + \right. \\ \left. + 3a_3 \omega H_{dc} H_{exc.m}^2 \sin(2\omega t) + \frac{3a_3 \omega H_{exc.m}^3 \cos(3\omega t)}{4}\right] s w_2 \tag{9}$$

It follows from (9), that the useful component proportional to the measured flux density in a constant magnetic field, is on the excitation frequency ω and doubled excitation frequency 2ω. At the same time, the output signal is noisy on 3ω frequency. Therefore, a complex devise based on band-pass filters or synchronous detectors is required to detect the useful component in the output signal of the single-rod fluxgate. For example, methods of synchronous detection of the fluxgate maximum sensitivity imply the use of device with four synchronous detectors. The first two detectors sense in-phase and quadrature components of the output signal on the excitation frequency ω, while the second two sense them on the doubled excitation frequency 2ω.

For the reasons outlined earlier, single-rod fluxgates are not applied in practice. A differential fluxgate both with open (two-rod) and closed (toroid) cores has become widespread [6-7].

3 Differential fluxgate

Schematic view of the differential fluxgate with open core are given in figure 2.

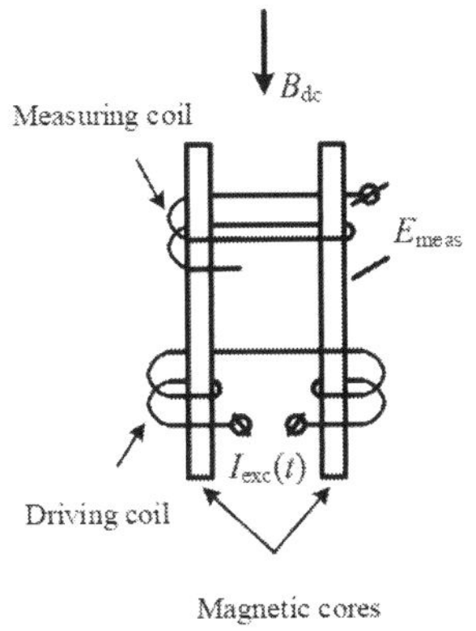

Fig.2. Differential fluxgate.

Let us consider the operation of the differential fluxgate with open core during measurement of the constant magnetic field. This fluxgate incorporates two driving coils with anti-parallel connection. The excitation current $I_{exc}(t)$ in them creates the magnetic intensity or excitation fields $H_{exc}(t)$ in each core. These excitation fields are characterized by the identical amplitude and phase opposition. In presence of the constant field H_{dc} directed along the core as shown in figure 2, superpositions of the excitation fields in the core bulk are calculated as:

$$H_{\Sigma 1}(t) = H_{dc} + H_{exc}(t);$$
$$H_{\Sigma 2}(t) = H_{dc} - H_{exc}(t), \quad (10)$$

Let us assume that the cores are identical. Their flux density is then obtained through the following equation, anisotropy and hysteresis being neglected:

$$B_1(t) = f\left(H_{\Sigma 1}(t)\right) = f\left(H_{dc} + H_{exc}(t)\right);$$
$$B_2(t) = f\left(H_{\Sigma 2}(t)\right) = f\left(H_{dc} - H_{exc}(t)\right). \quad (11)$$

The electromotive force induced in the measuring coil which, according to figure 2, envelopes both cores, is defined as

$$E_{mes}(t) = -sw_2\left(\frac{d}{dt}(B_1(t) + B_2(t))\right), \quad (12)$$

Using (10) and (6), we obtain the equation, which describes the flux density:

$$B_1(t) = a_1 H_{dc} + a_1 H_{exc}(t) + a_3 H_{dc}^3 + 3a_3 H_{dc}^2 H_{exc}(t) + 3a_3 H_{dc} H_{exc}^2(t) + a_3 H_{exc}^3(t);$$
$$B_2(t) = a_1 H_{dc} - a_1 H_{exc}(t) + a_3 H_{dc}^3 - 3a_3 H_{dc}^2 H_{exc}(t) + 3a_3 H_{dc} H_{exc}^2(t) - a_3 H_{exc}^3(t). \quad (13)$$

In the case of the fluxgate excitation by the sinusoidal alternating current, the substitution of (13) in (12) in view of (8) allows us to derive the electromotive force in the measuring coil:

$$E_{mes}(t) = 6sw_2 a_3 \omega H_{dc} H_{exc.m}^2 \sin(2\omega t). \quad (14)$$

So, unlike the single-rod fluxgate, the output signal of the differential fluxgate contains a useful component proportional to the measured flux density in a constant magnetic field just on the doubled excitation frequency 2ω.

The traditional analysis of the fluxgate operation which uses a third-order polynomial as in (6), is rather simplified and qualitative similar to a grapho-analytical method. The quantitative analysis is provided by the approximation of B/H dependence by higher order polynomials.

Let us analyze the differential fluxgate operation using the following approximation of B/H dependence:

$$B = a_1 H + a_3 H^3 + a_5 H^5. \qquad (15)$$

In the case of the fluxgate excitation by the sinusoidal alternating current, we insert (15) in (12). And in view of (11) and (8), the electromotive force in the measuring coil is found from

$$E_{\text{mes}}(t) = s w_2 \omega\left[-2H_{\text{dc}} H^2_{\text{exc.m}}\left(3a_3 + 10a_5 H^2_{\text{dc}} + 5a_5 H^2_{\text{exc.m}}\right)\sin(2\omega t) + 5a_5 H_{\text{dc}} H^4_{\text{exc.m}} \sin(4\omega t)\right] \qquad (16)$$

We can now proceed analogously to (16) when using the approximating dependence of the seventh order:

$$B = a_1 H + a_3 H^3 + a_5 H^5 + a_7 H^7, \qquad (17)$$

and thus obtain the electromotive force in the measuring coil:

$$\begin{aligned} E_{\text{meas}}(t) = s w_2 \omega \Bigg[& -\frac{1}{8} H_{\text{dc}} H^2_{\text{exc.m}} \left(48a_3 + 160a_5 H^2_{\text{dc}} + \right. \\ & \left. +80a_5 H^2_{\text{exc.m}} + 336a_7 H^4_{\text{dc}} + 105a_7 H^4_{\text{exc.m}} + 560a_7 H^2_{\text{dc}} H^2_{\text{exc.m}}\right)\sin(2\omega t) + \\ & +\frac{1}{2} H_{\text{dc}} H^4_{\text{exc.m}}(10a_5 + 70a_7 H^2_{\text{dc}} + 21a_7 H^2_{\text{exc.m}})\sin(4\omega t) + \frac{-21}{8} a_7 H_{\text{dc}} H^6_{\text{exc.m}} \sin(6\omega t) \Bigg]. \end{aligned} \qquad (18)$$

As can be seen from (16) and (18), the useful component proportional to the constant magnetic field measured, is observed in the output signal not only at the doubled excitation frequency 2ω, but also at 4ω and 6ω frequencies.

The analysis of (16) and (18) allows us to propose the procedure for processing signals from the fluxgate sensor in order to enhance the fluxgate sensitivity due to measuring signals not only at the doubled excitation frequency but also at other even frequencies with the subsequent summing the results.

In practice, not only sinusoidal but also rectangular and triangular waveforms are used in the capacity of a drive signal of fluxgate magnetometers. So, it is advisable to derive a general equation for computing the electromotive force in the measuring coil independent of the waveform. Let us write the equation for the magnetic intensity of exciting field as a Fourier series without the constant component:

$$H_{\text{exc}}(t) = H_{\text{exc.m}} \sum_{g=1}^{m}\left(b_g \cos(g\omega t) + c_g \sin(g\omega t)\right), \qquad (19)$$

where g is the number of harmonics in the drive signal; b_g , c_g are Fourier serial expansion coefficients.

Substituting (5) into (12) in view of (11) and 19), we obtain the resulting statement for the electromotive force:

$$E_{\text{mes}}(t) = -sw_2 H_q(t)\sum_{i=1}^{j} a_{2i-1}(2i-1)\left[\left(H_{\text{dc}}+H_{\text{exc}}(t)\right)^{2i-2} - \left(H_{\text{dc}}-H_{\text{exc}}(t)\right)^{2i-2}\right];$$
$$H_q(t) = \frac{dH_{\text{exc}}(t)}{dt} = \omega H_{\text{exc.m}} \sum_{g=1}^{m}\left(b_g \cos(g\omega t + \pi/2) + c_g \sin(g\omega t + \pi/2)\right)g. \qquad (20)$$

Now we can derive the differential fluxgate sensitivity at any harmonics according to (20):

$$S_g = \lim_{T\to\infty}\frac{1}{ET}\int_0^T \frac{dE_{\text{mes}}(t)}{dH_{\text{dc}}}\sin(g\omega t)dt =$$
$$= \lim_{T\to\infty}\frac{1}{ET}\int_0^T -sw_2 H_q(t)\sum_{i=1}^{j} a_{2i-1}(2i-1)(2i-2)\left[\left(H_{\text{dc}}+H_{\text{exc}}(t)\right)^{2i-3} - \right. \qquad (21)$$
$$\left. -\left(H_{\text{dc}}-H_{\text{exc}}(t)\right)^{2i-3}\right]\sin(g\omega t)dt.$$

This equation facilitates the analytical computation of the differential fluxgate sensitivity independent of the frequency content of the drive signal at a given approximation degree for $B = f(H)$ function.

4 Conclusions

The obtained analytical equations allowed us to calculate the electromotive force in the measuring coil of the fluxgate magnetometer independent of the drive signal frequency content. Also, the obtained equations provided the estimation of the fluxgate sensitivity at any harmonic and were used to study its operation with a glance to the waveform and the polynomial approximation of the mean magnetization curve of the core.

This research was conducted in Tomsk Polytechnic University and the research of operating principle of a single-rod fluxgate financially supported by the Ministry of Education and Science of the Russian Federation (agreement No.14.578.21.0232, unique identifier RFMEFI57817X232). The research of differential fluxgate model is funded from Russian Science Foundation (RSF), Grant Number 17-79-10083.

References

1. P. Ripka, IEEE Sensors journal **10**, 1108 (2010)
2. S. V. Uchaikin, *LT21*, 2809 (1996)
3. J. E. Lenz, IEEE Sensors journal **78**, 631 (1990)
4. P. Ripka, *Magnetic Sensors and Magnetometers* (Artech house, 2000)
5. V. E. Baranova, et al. *XXI IMEKO World Congress*, (2015)
6. Baschirotto, Measurement **43**, 46 (2010)
7. Hava Can, Uğur Topal, J Supercond Nov Magn **28**, 1093 (2015)

Torque motor tape winding characteristics and its connection with design parameters

Antonina Dolgih[*], and *Vladimir* Martemyanov

Tomsk Polytechnic University, 634050 Tomsk, Russia

Abstract. The paper considers the features of a tape winding torque motor development. Novel motor type allows to obtain an increased torque by the high current flowing through the tape winding. It is shown that the most effective way of the torque increasing is the tape winding plate resistance reducing.

1 Introduction

At present, torque motors are used in technical systems as the gearless drive base. The torque motor has a modular structure, consists of two separate assembly units: a rotor and a stator. The modular motor structure is used for direct integration into the mechanism. The rotor contains high-coercive rare-earth magnets. The stator has two design versions: "a smooth stator" and "a grooving stator". For precise systems "the smooth stator" is preferable, because there is no cogging torque.

The current state of the torque motors development is performed in [1-4]. The analysis shows that the specific characteristics of Russian torque motors are worse than foreign analogues. Torque motors can be improved by using new magnetic materials, changing the rotor design or using new types of windings. In our opinion, the most promising way is the third approach. In this case the specific characteristics increasing can be achieved by the torque raising due to the high current flowing through the rotor winding. In this case, there are some problems with the motor thermal mode, because the power consumption is mainly released as heat in the windings.

In this situation, it is proposed to replace the traditional winding of the engine with a tape winding with side cutouts. The heat in the winding will be directly transferred to the motor case, and the developed torque increasing will occur due to the high current flowing through the tape winding. In addition, a new motor stator design and technological approach are proposed, including the use of additive technologies; and the possibility of creating the necessary "torque-rotor angle" characteristic. Figure 1 shows the scheme of the tape winding torque motor.

[*] Corresponding author: ivanovatonya@tpu.ru

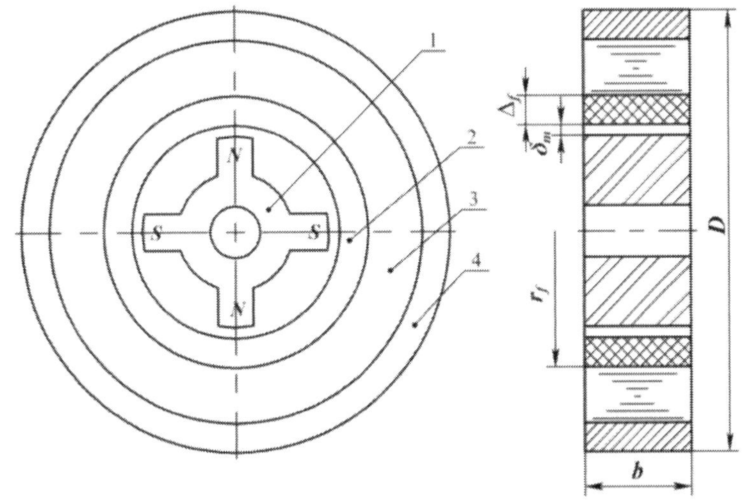

Fig. 1. The scheme of the tape winding torque motor:
1 - Permanent magnets (PM); 2 - Tube frame; 3 - Tape winding; 4 - Core.

The aim of this paper is the determination of the torque dependence over the design parameters of the motor, primarily geometric.

2 The motor's torque

It was previously shown [5] that the tape winding motor torque is given by the expression:

$$T = \frac{\Delta^2}{2} \cdot \frac{U}{R} \cdot B(\delta) \left[N + \frac{2r_f}{\Delta} - 1 \right] \cdot \frac{D(x, y, I_0)}{I_0}, \tag{1}$$

where Δ is the tape thickness; U is the power supply voltage; R is the plate resistance; $B(\delta)$ is magnetic flux density created by one pole; N is the number of winding turns; r_f is the tube frame radius; $D(x, y, I_0)$ is a double integral over the surface of the magnetic flux action.

In the above expression it is assumed that the thickness of the electrical insulation is small and neglected. Also, the expression (1) does not reflect the armature reaction on magnetic flux density. In addition, the component $\frac{2r_f}{\Delta} >> 1$ and the expression (1) can be reduced to the form:

$$T = \frac{U}{R} \cdot \left(\frac{N \cdot \Delta}{2} + r_f \right) \cdot \left[\frac{B(\delta)}{I_0} \cdot \Delta \cdot D(x, y, I_0) \right]. \tag{2}$$

Note that the expression in the square brackets represents the force produced by a single plate with 1 A current flowing through it:

$$F_{01} = \frac{B(\delta)}{I_0} \cdot \Delta \cdot D(x, y, I_0). \tag{3}$$

The expression $\left(\frac{N \cdot \Delta}{2} + r_f\right)$ represents the medium radius r_m of the tape winding. This allows us to indicate that:

$$T = \frac{U}{R} \cdot r_m \cdot F_{01}. \tag{4}$$

3 Results and discussion

However it is necessary to take into account the insulation on the tape surface, which can be done with the copper space factor:

$$k_c = \frac{\Delta}{\Delta_{in}}, \tag{5}$$

where Δ_{in} is the tape thickness with the insulation.

Then the expression for the medium radius will be presented as:

$$r_m = \frac{N \cdot \Delta}{2k_c} + r_0. \tag{6}$$

The flux density $B(\delta)$ and the integral $D(x, y, I_0)$ will also depend on the motor geometric parameters. The flux density mainly depends on the value of the non-magnetic gap δ_{nm}, which is determined by the sum of the mechanical gap δ_m, the thickness of the tube frame for winding the tape Δ_f and the tape winding thickness:

$$\delta_{nm} = \delta_m + \Delta_f + \frac{\Delta \cdot N}{k_c}. \tag{7}$$

Besides, as shown previously [6-8], the magnetic flux density also depends on the pole pairs number of the magnetic system. Also in [6] it is indicated that for a given tape width the value of the double integral $D(x, y, I_0)$ remains practically constant when the pairs of poles number of the magnetic system changes.

As a result, it is possible to estimate the influence of the force value F_{01} at the motor's torque only by analyzing the particular motor scheme.

Let us consider the influence of the single plate resistance on the motor's torque. Resistance depends on the motor geometric parameters. As the poles number of the magnetic system increases, the plate length corresponding to the pole division will decrease. The plate resistance over its length will vary ambiguously. Numerical simulations have shown that the resistance of a single plate depends on the ratio of its length a and width b. Varying this ratio $\beta = a/b$, when b=const, a certain minimum resistance value is observed (Figure 2). During the modeling in the Electric Currents module of COMSOL, a copper plate with thickness $\Delta = 10^{-4}$ m, a width b = 0.05 m was constructed. The plate length a_i varied from the initial value $a_1 = 0.135$ m.

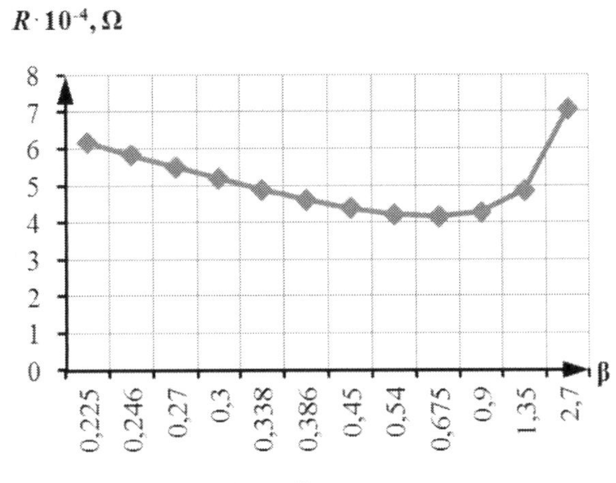

Fig. 2. The single plate resistance over ratio $\beta = a/b$.

The average plate length (pole pitch) is defined as

$$a = \pi \cdot \frac{r_m}{p}, \tag{8}$$

where p is a number of magnetic system pole-pairs.

The tape width is almost equal to the motor axial length b. The minimum plate resistance R_{min} is reached at a specific value of the ratio β. If we substitute (8) in the formula for the ratio β, we obtain:

$$\beta = \frac{\pi}{p} \cdot \frac{r_m}{b}. \tag{9}$$

The necessary value of β can be obtained by:

1. Changing the medium radius of the tape winding due to the number of winding turns selection N;
2. Selecting the pairs of poles number p;
3. Changing the tape width b.

The most rational approach is the third, since the first two require separate studies.

4 Conclusion

Developing the torque motor with the high specific characteristics it is necessary to consider the influence of the design parameters of both the magnetic circuit and windings. Firstly the poles pairs number that provides the maximum flux density in the gap should be determined. The plate length a is chosen equal to the pole pitch of the magnetic circuit. The minimum value of β when the plate resistance is also minimum is selected by changing the tape width. Then the plate resistance value is calculated and, according to the permitted winding current, the required number of turns are founded.

References

1. G. Dubrovskiy, A. Mikerov, V. Dzhankhotov and J. Pyrhonen, EPE-ECCE Europe 6910754 (2014)

2. D. Hanselman, *Brushless Permanent Magnet Motor Design. Second Edition* (Magna Physics Publishing, United States of America, 2006)
3. A. Hughes and B. Drury, *Electric Motors and Drives. Fundamentals, Types and Applications. 4th Edition* (Elsevier-Newnes, Oxford, 2013)
4. P. Madaan, *Brushless DC Motors. Part 1. Construction and Operating Principles* (Cypress Semiconductor, India, 2013)
5. A. Dolgih (Ivanova), V. Martemyanov, *SIBCON,* 7491844 (2016) DOI: 10.1109/SIBCON.2016.7491844
6. A. Dolgih (Ivanova), V. Martemyanov, V. Borikov, Matec Web of Conf. 113, 01013 (2017) DOI: 10.1051/matecconf/201711301013
7. N. Natalinova, N. Ilina, E. Frantcuzskaia, IOP Conf. Ser.: Materials Science and Engineering **132(1)**, 012029 (2016) DOI: 10.1088/1757-899X/132/1/012029
8. I. A. Lezhnina, A. A. Boyakhchyan, K. V. Overchuk, A. A. Uvarov, Journal of Physics: Conf. Ser. **881(1)**, 012041 (2017) DOI: 10.1088/1742-6596/881/1/012041

Laboratory testing of LoRa modulation for CubeSat radio communications

Alexander Doroshkin[1], *Alexander* Zadorozhny[1,*], *Oleg* Kus[2], *Vitaliy* Prokopyev[1], and *Yuri* Prokopyev[1]

[1]Novosibirsk State University, 630090 Novosibirsk, Russia
[2]OKB Fifth Generation Ltd., 630090 Novosibirsk, Russia

Abstract. We report the results of the feasibility study of using LoRa modulation for radio communications between CubeSat at low Earth orbit and ground station. The main goal of the study is to define how Doppler effect affects a LoRa radio link. Results of laboratory testing have shown high immunity of LoRa radio link to Doppler shift and a possibility to use LoRa modulation in CubeSat radio communications without any limitations.

1 Introduction

Various modifications of frequency or phase modulation are usually applied in the radio communication systems of the nanosatellites CubeSat. However, using more sophisticated modulation methods can significantly advance the efficiency of the radio link. One of such methods is the LoRa modulation [1-3] widely used in the networks of machine-to-machine (M2M) communications in the networks of IoT wireless sensors. This modulation method is based on the spread-spectrum technique where the data is encoded by a wide-band chirp signal in which the frequency linear increases or decreases with time.

A peculiarity of the Earth-satellite radio link is a high speed of the spacecraft passing over the ground station, which causes a strong Doppler shift of the frequency of the transmitted signal at the receiving point. The spread-spectrum technique used in the LoRa modulation has an inherent low sensitivity to the Doppler effect [2]. However, the LoRa modulation specification does not contain clear criteria of its applicability under real conditions of a quick movement of receiver relatively transmitter. It concerns especially conditions with the dynamic Doppler effect when the frequency shift changes rapidly with time. This takes a place in the Earth-satellite radio link. Therefore, it is yet unclear whether it is possible to use the LoRa modulation in radio link between a satellite and ground station. This paper presents the results of laboratory study of the immunity of a radio link with the LoRa modulation to the Doppler effect under the conditions corresponding to low Earth orbit CubeSat flying over the ground radio communication station.

* Corresponding author: zadorozh@phys.nsu.ru

2 Laboratory equipment for testing LoRa modulation

Radio frequency vector signal generator NI PXIe-5672 by National Instruments and a LoRa transceiver based on the SX1278 chip by Semtech Corporation were used for testing and experimental studying the radio link using the LoRa modulation.

The NI PXIe-5672 vector signal generator is used for the synthesis of a RF signal with the LoRa modulation. In addition, the NI PXIe-5672 makes it possible to introduce various random and deterministic interference into the synthesized signal, including frequency distortions caused by the Doppler effect during the passing of the satellite over a ground station. The LoRa signal is synthesized by means of the software, which was developed for this research in the LabVIEW using the NI-RFSG software. In the experiments performed, the LoRa radio signal with a carrier frequency of 430 MHz was synthesized.

The LoRa SX1278 standard transceiver is used during the laboratory experiments as a receiver and a demodulator. The LoRa radio signal emitted by the NI PXIe-5672 generator, being synthesized taking into account Doppler shift, is received, demodulated and analyzed by the receiver. The analysis of the quality of the received signal and the data packages contained therein are carried out using data from the internal registers of the receiver.

3 Testing of LoRa radio link for immunity to Doppler shift

In order to synthesize the high-frequency radio signal with the LoRa modulation propagating in a radio link between the satellite and the ground station, it is necessary to set the Doppler frequency shift and its variation in time while the satellite is flying over the station. Let the transmitter emits a radio signal with a frequency F_0. Due to the Doppler effect, the receiver will get a signal with frequency

$$F = \frac{1}{1 + \frac{v}{c}\cos\beta} \cdot F_0 \quad (1)$$

where v is the satellite velocity, c is the light speed, β is the angle between the satellite velocity vector and the direction to the ground station. Then the relative Doppler frequency shift δF is defined by:

$$\delta F = \frac{F - F_0}{F_0} = \frac{1}{1 + \frac{v}{c} \cdot \cos(\beta)} - 1 \,. \quad (2)$$

Let us consider the satellite moving in a circular orbit of height H and ground station situated in a plain of the orbit (Figure 1).

Taking into account that the satellite velocity in the circular orbit $v = \sqrt{g \cdot R / (1 + H / R)}$ where R is the Earth's radius and g is the gravitational acceleration at the Earth's surface, it is easy to obtain the following from (2) and Figure 1:

$$\delta F = \frac{1}{1 + \frac{1}{c} \cdot \sqrt{\frac{g \cdot R}{1 + H / R}} \cdot \frac{\sin(\varphi)}{\sqrt{(1 + H / R)^2 - 2 \cdot (1 + H / R) \cdot \cos(\varphi) + 1}}} - 1 \quad (3)$$

where

$$\varphi = \frac{\sqrt{g/R}}{\left(1+H/R\right)^{3/2}} \cdot t \qquad (4)$$

Here, the time t is assumed to equal zero when the satellite is at the zenith above the ground station.

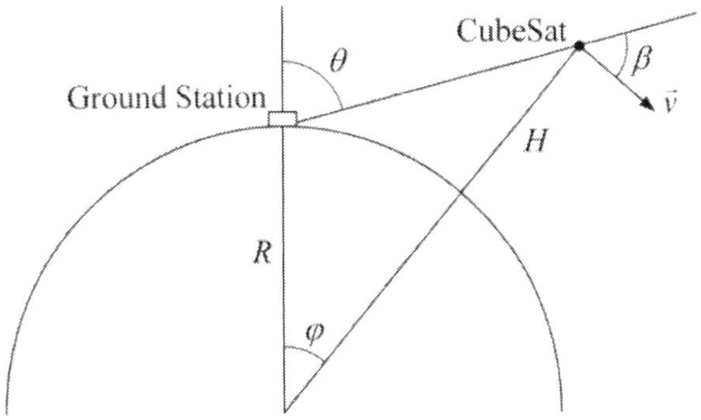

Fig. 1. The motion of the satellite in a circular orbit with respect to the ground station.

The Doppler effect is the most pronounced when the satellite moves at low altitudes, therefore its influence on the radio link with the LoRa modulation was investigated for the satellite moving in a circular orbit of 200 km high. The relative Doppler shift δF in ppm units (parts per million) calculated by the formula (3) for this orbit and its time derivative are shown in Figure 2.

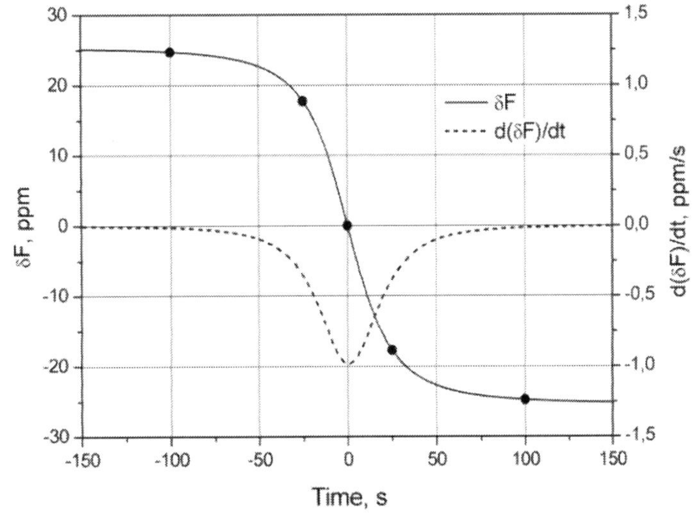

Fig.2. The relative Doppler shift δF and its time derivative during the passing of the satellite at 200 km height over the ground station. The satellite is at the zenith at the time t = 0. Bold points show times for which laboratory measurements were carried out.

In our experiments, the NI PXIe-5672 vector signal generator synthesized a RF LoRa signal with a frequency varying due to the Doppler effect in accordance with Figure 2. That is the synthesized RF signal takes into account the dynamic Doppler Effect – the changing with time Doppler shift. The Doppler shift change reaches its maximum value when the satellite is at the zenith over the ground station. The synthesized RF signal was applied to the input of the SX1278 transceiver operating as a receiver and a demodulator. The criterion of communication quality was a 100% number of data received without loss from the entire package of the transmitted data. The data loss was detected via the check sum. The measurements were made at various LoRa modulation parameters and package durations.

The main parameters of the LoRa modulation, which determine the bit rate of the data signal, are the spreading factor (*SF*) and the spread spectrum modulation bandwidth (*BW*). The SX1278 transceiver using in the experiments offers *BW* options ranging from 7.8 kHz to 500 kHz with *SF* ranging from 6 to 12. The immunity of the radio link with the LoRa modulation to the Doppler effect was tested with two sets of the parameters, which are given in Table 1. In the first set, the transmission time of a LoRa packet is close to the lowest possible in the standard LoRa for maximum payload size, and in the second set, it is close to the maximum possible one.

Table 1. LoRa modulation parameters used in testing LoRa radio link for immunity to Doppler effect.

No. of set	*SF*	*BW*, ppm	LoRa payload size, byte	Transmission time of a LoRa packet, s
1	7	125	255	0.55
2	11	250	255	2.5

Bold points in Figure 2 show the most specific elements of satellite trajectory in the context of the radio link conditions covering all extreme situations. These situations are the largest static Doppler effect when the satellite is near the horizon, the largest dynamic Doppler effect when the satellite is at the zenith, and simultaneous high Doppler shift and high velocity of its changes when the satellite is in the intermediate position. Testing LoRa radio link was carried out for these satellite trajectory legs.

To determine the LoRa radio link immunity margin at each selected elements of satellite trajectory, the LoRa modulated radio link was tested for several δF values and its derivative. First, tests were carried out at nominal values, which are calculated by the formula (3) and are shown in Figure 2. Then the Doppler shift was multiplied by the escalation factor $Q > 1$, and the tests were repeated for various increasing values of Q until the connection was lost. In such a manner, the maximum possible value Q_{max} characterizing the LoRa radio link immunity margin to the Doppler effect was determined.

Preliminary measurements showed that an increase in the RF signal level did not cause the increase in the number of successfully received packages unless the Doppler shift became more critical one, and the connection was lost. Therefore, the level of RF LoRa signal was chosen slightly above the receiver's sensitivity threshold in all the experiments. The results of the radio link tests with LoRa modulation parameters from Table 1 are given in Table 2. Here δF_{max} and $(d(\delta F)/dt)_{max}$ are the maximum values of the relative Doppler shift and its time derivative under which the data transmitted without loss, Q_{max} is corresponding immunity margin to the Doppler effect.

Table 2. Results of testing LoRa radio link for immunity to Doppler effect.

SF, *BW*	Static Doppler effect (t = 100 s)		Dynamic Doppler effect (t = 0 s)	
	Q_{max}	δF_{max}, ppm	Q_{max}	$(d(\delta F)/dt)_{max}$, ppm/s
7, 125	2.8	70	>10	>10
11, 250	5.2	130	4	3.9

One can see from Table 2 that the tested LoRa radio link has, in the worst case, the immunity margin to the Doppler effect Q_{max} = 2.8. That is, in the worst case, the Doppler effect can disrupt the LoRa radio link performance only when the Doppler shift is more than 2.8 times greater than the shift caused by the satellite moving in a circular orbit of 200 km high. It is important to note that the LoRa radio link is more immune to the dynamic Doppler effect in case of the first set of the LoRa modulation parameters (SF=7, BW=125 kHz) that is in case of the shorter transmission time of a LoRa packet. Under longer transmission time of a LoRa packet (SF = 11, BW = 250), the situation is reversed. In the case of the static Doppler effect, the measured values δF_{max} (70 and 130 ppm) coincide rather well with the specification of the LoRa SX1278 transceiver [4] in the part of the tolerated frequency offset between the transmitter and the receiver for the same modulation modes and the carrier frequency of 430 MHz (73 and 100 ppm, respectively). As for the dynamic Doppler effect, it should be noted that the immunity of the LoRa radio link under the considered conditions is ensured by the SX1278 transceiver only in the mode which supported the symbol synchronization. The third bit of the *RegModemConfig3* register activates this mode.

4 Conclusion

The performed laboratory study showed a high immunity of the radio link using the LoRa modulation to the Doppler effect under conditions of radio communication between satellite at low Earth orbit and ground station. The immunity margin to the Doppler effect under these conditions is better than 2.8. The impact on the LoRa radio link of both static and dynamic Doppler effect was investigated. It is obtained that the immunity to the dynamic Doppler effect is ensured only in the mode which supported the symbol synchronization. In general, the obtained results showed that the LoRa modulation could be used in low Earth orbit CubeSat radio communications without any limitations.

This work was supported by the Ministry of Education and Science of the Russian Federation: Project No. 14.575.21.0154, Project Identification No. RFMEFI57517X0154.

References

1. Fractional-N Synthesized Chirp Generator. United States Patent No.: US 7,791,415 B2, Date of Patent: Sep. 7 (2010)
2. *LoRa™ Modulation Basics. Application None AN1200.22* (Semtech Corporation, 2015)
3. L. Vangelista, IEEE Signal Processing Letters, **24**, 1818 (2017)
4. *SX1276/77/78/79 – 137 MHz to 1020 MHz Low Power Long Range Transceiver. Datasheet, Rev. 5* (Semtech Corporation, 2016)

Organization design of complex technical products in integrated information systems

Evgeny Dubrovsky[*], and *Viktor* Dmitriev

Tomsk Polytechnic University, 634050 Tomsk, Russian Federation

Abstract. The article substantiates the necessity of using modern tools for the design of advanced space technology. The problems and consequences are presented in the application of CAD of various profiles within the framework of a single technical project. The decision of problems is resulted. The description of the through design process and the integrated information system is given. The possibilities of integrated information systems and the expected business effect are shown in their implementation and organization of an end-to-end design process.

1 Introduction

Engineering of spacecraft is complex and hi-tech process. The most part of highly qualified engineers are working on the design of spacecraft. In variety of industries the market sees a lot of change happening these days; new trends and challenges obviously need to be addressed as part of the product creation process [1]. Engineering design is a source of competitive advantage for manufacturing companies [2-8]. Manufacturing finished products and testing engineers using special CAD\CAM\CAE systems are preparing production for design and modelling [9]. CAD\CAM\CAE systems allow to gain more speed in project development, and to reduce duration of development new spacecraft's, and to improve quality development because of speed and reliability of computer technology, reliability of mathematical models, effective methods of optimization, and exclusion of the human factor. Use of CAD\CAM\CAE systems allows more qualitatively to solving questions related to:
- A large amount of work on the design and modeling of products;
- Time and financial constraints in the design;
- The need for a detailed engineering of the spacecraft as a whole;
- The need to organize stage-by-stage control of the design and manufacture of the finished product.

2 CAD\CAM\CAE for industry

In this article CAD\CAM\CAE systems widely used in industry are considered. The most important points are:

[*] Corresponding author: evgenij.dubrovskij@gmail.com

- ECAD – system for electrical design;
- MCAD – system for part and product design.

2.1 ECAD and MCAD

Between ECAD and MCAD systems, there are cardinal differences in the functional purpose and data representation, but integrated using of ECAD and MCAD systems allows quick and qualitative problem solving:
- Preparation of input data;
- Electric preparation of production;
- Design preparation of production;
- Technological preparation of production.

Engineering of spacecraft requires a well-developed end-to-end design technology, where the design process should be closely related, and ECAD and MCAD integrated into a single information system. Let us explain in more detail.

2.2 Problems with integration

The design process of a complex technical product is divided into electrical and design parts, in order to meet both part's requirements, integrated use of the MCAD and ECAD systems in the same project is highly recommended. Usually data's getting from ECAD system (electrical circuits) are input data for MCAD systems. For example, block layout design, product design and cables network topology are made with using of data obtained from the development results of ECAD product schematic diagrams. At the same time, the task of electrical design is to organize the correct connection of wires, connectors, cables and the organization of the required signal transmission between the equipment. In turn, the task of designing the product design is to organize the maximum compactness of the arrangement of devices and the optimal topology of the cable network.

Designing a product by using of two or more different CAD systems imposes necessity of data exchange between them. That, in turn, requires development or implementation of special software with subsystems that ensure mutual integration, because the problem is that data cannot always be transferred between CAD without preliminary conversion. Converting data from one format to other often results in the accumulation of errors in the design process. Errors during design process will cause negative influence to design results, and it can be detected only during manufacturing, installation or testing process, what is absolutely unacceptable.

On practice, the problem is complicated by the fact that within a single project, specialists can use several ECAD systems that do not have direct data exchange with a 3D MCAD system.

Consequence of the lack of integration between ECAD and MCAD systems is the often inconsistencies between the electrical circuits and design of product. If there is an error in the electrical circuit or the designer can not implement the circuit design, there is no way to correct quickly the situation. Often it is necessary to repeat the stage of electrical design again, or to spent time for revision, which is by no means fast. In addition, during the revision process construction design information can become out of date. As the result, it is a great loss of time. Solving of this conflict situations is directly depends on human factor and it is not always the result of objective analysis, so it can't leads to the optimal decision making.

3 Solutions

Engineers from different countries work over solution of described problems. Today, the most effective, universally recognized and used in the world practice is the organization of end-to-end design process at enterprises and development or implementation of integrated information systems instead of individual CAD systems.

The integrated approach is able to design, maintain and produce other data, process and metadata management in the database, monitoring its development [3]. The end-to-end process is continuous sequence of actions, where each next stage must be based on the results of previous stage. At the same time, it must be ensured that the operations are carried out not only in a directly, but also in the opposite direction, which makes it possible to solve problems more effectively:

- Return of the received input data for revision of dimensional parameters;
- Operative finding of errors in the circuit solution;
- Revision of the electrical circuit;
- Ensuring maximum control over the life cycle of the product;
- Exclude the possibility of loss of design data;
- To ensure the security of data storage.

Integrated information systems represent the software for designing, built into a single information space based on a single software platform. In the framework of software platform, data can be directly exchanged between components, ECAD and MCAD, which must have good compatibility in all technological design operations.

Because of the fulfillment of this condition, the integrated information systems allow completely:

- To organize end-to-end design technology in practice;
- Maximize the design process;
- To differentiate the rights of access of developers;
- Reduce the costs of design, manufacture and testing;
- Reduce of errors the human factor, due to automation;
- Organize work in a unified system of accounting, control and storage of data;
- Promotes the organization of paperless design technology.

4 Conclusions

The organization of an end-to-end design and implementation of integrated information systems leads to:

- improving the quality of products;
- reducing the cost of products;
- Reducing time to market.

As a result, the solution of these problems leads to the main goal of the enterprise - to increase the competitiveness of the enterprise's products on the world market.

References

1. A. Barth, G.L. Kovács, D. Kochan, *IFIP AICT* **411** (2013)
2. T. Boudouh, S. Gomes, MATEC Web of Conferences **112** (2017)
3. M. G. Trotta, Journal of Industrial Engineering and Management (2010)
4. V. S Dmitriev, G. N. Gladushev, T. G.Kostuchenko, *KORUS* 23, (2001)
5. R. C. W. Sung, J. M. Ritchie, T. Lim , Z. Kosmadoudi, Virtual Reality (2012)

6. B. L. Wan, W. X. He, C. Y. Chen, MATEC Web of Conferences **44**, (2016)
7. A. V. Taracenko, V. S. Dmitriev, T. G.Kostuchenko, *KORUS* 83, (2005)
8. Yu. Britova, V. Dmitriev, T. Kostyuchenko, IOP Conference Series: Materials Science and Engineering **132(1)**, (2016)
9. V. E. Baranova, P. F. Baranov, Dynamics of Systems, Mechanisms and Machines (2014)

Real-time perspective correction in video stream

Vladislav Glagolev[1,*], and *Alexander* Ladonkin[1]

[1]Control Devices department, Tula State University, 300600, pr. Lenina, 92, Tula, Russia

Abstract. The paper describes an algorithm used for software perspective correction. The algorithm uses the camera's orientation angles and transforms the coordinates of pixels on a source image to coordinates on a virtual image form the camera whose focal plane is perpendicular to the gravity vector. This algorithm can be used as a low-cost replacement of a gyrostabilazer in specific applications that restrict using movable parts or heavy and pricey equipment.

1 Introduction

Various computer vision tasks may require a video camera images to satisfy a number of conditions. One of such conditions is a constantly horizontal orientation of the camera, which is especially relevant for video-based measurements. This means the camera should be always pointed down even if the object that holds it is leaned. The mentioned requirement can be satisfied using a gyrostabilizer, however, this solution implies fixing a camera on a movable platform, which greatly increases the system complexity and reduces its reliability.

Solving this problem in software seems to be a good option for wide-angle cameras. The general idea is to acquire the camera's orientation relative to the earth and wrap the perspective according to it.

2 Description

Consider an object that holds a video camera. Its orientation is defined by three angles: roll (γ), pitch (ϑ) and yaw (ψ) as shown on figure 1.

In the simplest case, the γ and ϑ angles can be determined using a micromechanical accelerometer, although the accelerometer-only measurements are the subject to significant errors due to linear acceleration. A complex attitude control system is a better option.

The γ and ϑ Euler angles determine the position of an object relative to the surface, which allows for transformation that compensate the perspective wrap.

The required transformation is performed with the matrix operation (1), which transforms raw pixel coordinates (x_0, y_0) into the (x, y) coordinates and compensates the perspective wrap.

*Corresponding author: tgupu@yandex.ru

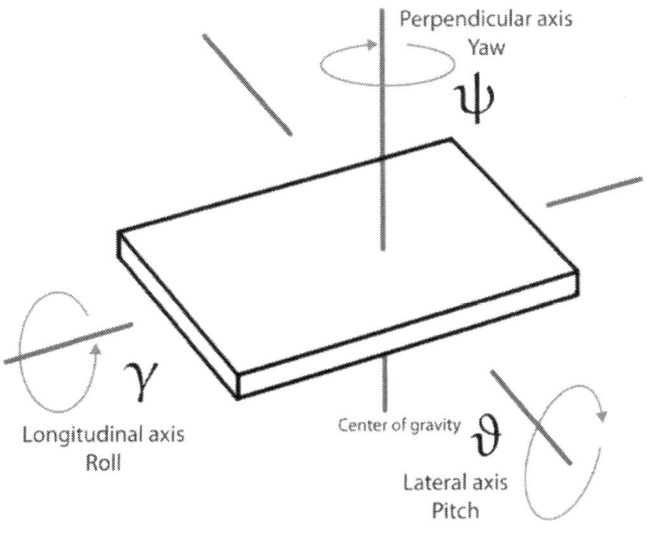

Fig. 1. Euler angles

$$\begin{pmatrix} y_1 \\ z_1 \\ x_1 \end{pmatrix} = \begin{pmatrix} \cos\gamma & \sin\gamma & 0 \\ -\sin\gamma & \cos\gamma & 0 \\ 0 & 0 & 1 \end{pmatrix} \times \begin{pmatrix} 1 & 0 & 0 \\ 0 & \cos\vartheta & \sin\vartheta \\ 0 & -\sin\vartheta & \cos\vartheta \end{pmatrix} \times \begin{pmatrix} y_0 \\ z_c \\ x_0 \end{pmatrix};$$

$$\begin{cases} x = x_1 \dfrac{f}{z_1}; \\ y = y_1 \dfrac{f}{z_1}, \end{cases} \tag{1}$$

where: z_c depends on camera's viewing angle and should be treated as a **response** of the perspective wrapper on angle values;

f—**scaling coefficient** that normally equals to the lens' focal length, but may be changed for scaling purposes.

The formula assumes that the origin of coordinates is located in the optical center (c_x, c_y) and the image is not distorted. These requirements can be achieved with the camera calibration.

The exact values of z_c and f parameters are different for each camera and can be determined experimentally, because the analytic expression $z_c = \dfrac{c_x}{\tan(\omega_{c_x})}$ includes the camera's viewing angle ω_{c_x} which is also an experimental value (in general case).

The rotation for a yaw angle is missing because the yaw stabilization is not a common task. If you need to stabilize the ψ angle, multiply the source vector $(y_0, z_c, x_0)^T$ by the matrix (2) before applying the formula (1).

$$\begin{pmatrix} \cos\psi & 0 & \sin\psi \\ 0 & 1 & 0 \\ -\sin\psi & 0 & \cos\psi \end{pmatrix} \tag{2}$$

The described method enables you to recalculate the points of interest coordinates or change the image by wrapping the perspective based on the four angle points' coordinates.

3 Implementation

In order to test the described method, a smartphone application was developed. Any modern smartphone has accelerometer and camera, which makes them a great platform for testing such methods. The application was created for an Android smartphone using the Kotlin language.

The OpenCV library was used to simplify image transformations and matrix operations. This library also implements camera calibration method [1], which was used to find the *Camera Matrix* (A) and *Distortion Coefficients* (D) (3) according to the Brown–Conrady's camera model [2, 3]:

$$A = \begin{pmatrix} f_x & 0 & c_x \\ 0 & f_y & c_y \\ 0 & 0 & 1 \end{pmatrix}; \quad D = \begin{pmatrix} k_1 \\ k_2 \\ p_1 \\ p_2 \\ k_3 \end{pmatrix}. \tag{3}$$

The Kotlin implementation is an **UnTilter** class whose main methods are **remapPoint**, **remapFrame** and **unTilt**:

```
fun remapPoint(point: Point, pitch: Double, roll: Double): Point{
  val (sinGamma, cosGamma) = Pair(Math.sin(roll),  Math.cos(roll))  // x
  val (sinTheta, cosTheta) = Pair(Math.sin(pitch), Math.cos(pitch)) // y

  val G = Mat(3, 3, CvType.CV_32F)  // gamma transformation matrix
  G.put(0, 0,  cosGamma);  G.put(0, 1,  sinGamma);  G.put(0, 2, 0.0)
  G.put(1, 0, -sinGamma);  G.put(1, 1,  cosGamma);  G.put(1, 2, 0.0)
  G.put(2, 0,0.0);  G.put(2, 1,0.0);  G.put(2, 2, 1.0)

  val T = Mat(3, 3, CvType.CV_32F)  // theta transformation matrix
  T.put(0, 0, 1.0);  T.put(0, 1,0.0);  T.put(0, 2,0.0)
  T.put(1, 0, 0.0);  T.put(1, 1,  cosTheta);  T.put(1, 2, -sinTheta)
  T.put(2, 0, 0.0);  T.put(2, 1,  sinTheta);  T.put(2, 2,  cosTheta)

  val A = Mat(3, 1, CvType.CV_32F)  // source vector
  A.put(0, 0, point.y);  A.put(1, 0, zc);  A.put(2, 0, point.x)

  // 1.psi  2.theta  3.gamma: A = G . (T . (P . A))
  Core.gemm(G, T, 1.0, Mat(), 0.0, G)  // G = G . T
  Core.gemm(G, A, 1.0, Mat(), 0.0, A)  // A = G . A

  val z1 = A.get(1, 0)[0]
  return Point(A.get(2, 0)[0] * f / z1, A.get(0, 0)[0] * f / z1)
}

fun remapFrame(pitch: Double, roll: Double) {
  val c = Point(cameraMatrix.get(0, 2)[0], cameraMatrix.get(1, 2)[0])
  var p = Point(0.0, 0.0)
  for (i in 0..3) {  // Calculating in four corners
    when (i) {
      0 -> { p.x = -c.x;            p.y = -c.y            } //top left
      1 -> { p.x = -c.x;            p.y = imgHeight - c.y } //bottom left
      2 -> { p.x = imgWidth - c.x; p.y = imgHeight - c.y } //bottom right
      3 -> { p.x = imgWidth - c.x; p.y = -c.y            } //top right
    }
    p = remapPoint(p, pitch, roll)
    p.x += c.x;  p.y += c.y
    outputQuad.put(i, 0, p.x, p.y)
  }
}
```

```
fun unTilt(img: Mat, undistort: Boolean = false){
  if (undistort) Imgproc.remap(img, img, undistortMap1, undistortMap2, 1)
  Imgproc.warpPerspective(img, img,
              Imgproc.getPerspectiveTransform(inputQuad, outputQuad),
              Size(img.width().toDouble(), img.height().toDouble()))
}
```

Note that the sinuses in matrix transformation for the ϑ angle are inverted. The reason for that is the difference between the positive direction of the axis assumed in the **`Sensor.TYPE_ROTATION_VECTOR`** Android's built-in virtual sensor and the direction assumed on figure 1.

The **`inputQuad`** and **`outputQuad`** matrices initialized as **`Mat(4, 1, CvType.CV_32FC2)`** are filled with the following values in the constructor:

```
inputQuad.put(0, 0,      0.0,       0.0) //top left
inputQuad.put(1, 0,      0.0, imgHeight) //bottom left
inputQuad.put(2, 0, imgWidth, imgHeight) //bottom right
inputQuad.put(3, 0, imgWidth,       0.0) //top right
inputQuad.copyTo(outputQuad)  // No wrapping by default
```

4 Testing

The presented code was tested on a **Sony XPERIA mini ST15i** smartphone. The figure 2 shows the screenshots taken during the testing with different angles. To make the stabilization clearer, three images from figure 2 were overlapped, the result is shown on figure 3. Sample shots were taken by hands, thus there were linear movements of the device between shots. These movements were compensated with the operations of translation, scaling, and rotation.

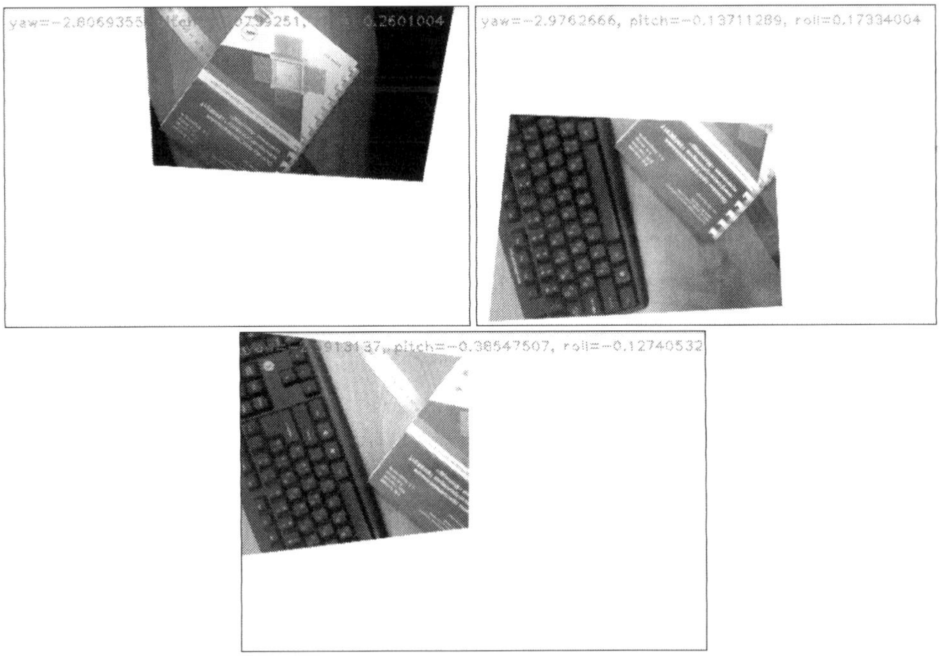

Fig. 2. Screenshots

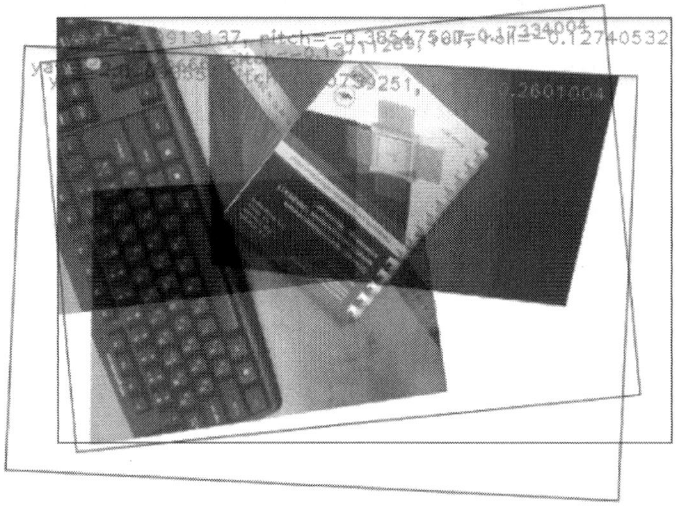

Fig. 3. The combination of three images acquired by multiplication of the images with 80% opacity

Objects remain approximately at the same place on the image regardless of the smartphone's tilt. The accuracy of tilt compensation depends on camera calibration quality, the orientation angles' precision and the precision of the **zc** parameter.

5 Conclusion

The described method is applicable when there is no need in the "untilted" image. For instance, some applications [4] require only the points of interest coordinates without the full image. In this case, the resulting coordinates can be calculated only for the points of interest and avoid the resource-demanding image transformations.

This also works for the undistortion procedure: it is possible to calculate the undistorted coordinates only for the points of interest without processing the entire image [5].

References

1. Z. Zhang, IEEE Trans. Pattern Anal. Mach. Intell. **22**, 1330 (2000)
2. D.C. Brown, Photometric Engineering **32**, 444 (1966)
3. A.E. Conrady, Monthly Notices of the Royal Astronomical Society **79**, 384 (1919)
4. V.M. Glagolev, A.V. Ladonkin, News of the Tula state university. Technical sciences **10**, 186 (2016)
5. V.M. Glagolev, News of the Tula state university. Technical sciences **9.2**, 188 (2017)

Foresight-audit of management's systems in aerospace engineering

Irina Gorelova[1,*], *Ekaterina* Harchevnikova[1], and *Maryam* Minigalieva[2]

[1]Russian Academy of national economy and public administration under the President of the Russian Federation, Volgograd Institute of management, 400131 Volgograd, Russia
[2]Ugra State University, Pedagogy Department, 628012 Khanty-Mansiysk, Khanty-Mansi Autonomous Area - Yugra, Russia

Abstract. Space engineering is an intensively developing sphere of human activity. The tasks of managing this sphere include the functions of designing and forecasting the future, as well as evaluating the predicted achievements and successes required for implementation, as well as possible disruptions and problems in the aerospace industry and resources. The evaluation of these resources acts as foresight audit. Foresight audit is a comprehensive internal and external assessment of the dynamic and static capabilities of aerospace companies. Foresight audit includes assessment of economic, social, organizational, psychological ideological and other aspects of the functioning and development of aerospace enterprises. The purpose of the study is the functions of designing and forecasting the future, as well as evaluating the predicted achievements and successes required for implementation, as well as possible disruptions and problems in the aerospace industry and resources. The cosmos will be "mastered" only with the mastering of oneself by man, only when harmony of man and the world is reached.

1 Introduction

Space engineering is an intensively developing sphere of human activity. Most important tasks of managing in the Space industry include the functions of designing and forecasting the future. Other tasks of managing in this sphere include the evaluating the predicted achievements and successes required for implementation, as well as the evaluating of the possible disruptions and problems in the aerospace industry and resources. The evaluation of these resources acts as foresight audit. Foresight audit is a comprehensive internal and external assessment of the dynamic and static capabilities of aerospace companies. Foresight audit includes assessment of economic, social, organizational, psychological ideological and other aspects of the functioning and development of aerospace enterprises. Leading aspects of foresight are aspects related to the study of the «human factor» of the current and predicted success of aerospace enterprises. To such aspects, it is possible to carry features and strategy of motivation of work. It is activity of the experts working at the enterprises of an aerospace branch and strategy of professional

* Corresponding author: mariam_ira@mail.ru

and career development of experts of branch as a whole. This can also include the features of management systems for enterprises in the aerospace industry, the notion of enterprises as organizations with mono-subjective (mono-actor) or intersubjective management (multi-actor). In addition, this is the view of the aerospace industry as a sphere of development and application of multi-agent technologies. Multi-agent technologies are the complex set of computer programs. Multi-agent technologies is included in the industry management system in general and in the particular aspects of this industry.

2 Materials and Methods

The primary method is theoretical integrative analysis. Of particular interest is the functions of designing and forecasting the future. Other tasks of managing in this sphere include the evaluating the predicted achievements and successes required for implementation, as well as the evaluating of the possible disruptions and problems in the aerospace industry and resources. The evaluation of these resources acts as foresight-audit. The authors discusses various aspects of the functions of designing and forecasting the future. The strategic management theory sees the source of competitive advantage in dynamic and static abilities, forecasting the future and the predicted achievements and successes required for implementation.

3 Results

The current problems of aerospace companies say that this industry is at the breaking point of its development. It needs an integrative foresight audit (multi-level and multi-component ore system audits) and audits of their current operations. The integrated foresight audit of aerospace companies is aimed at assessing the prospects and challenges, barriers and challenges of the industry development that are possible in the more or less distant future. Its purpose is to analyze the static and dynamic resources of enterprises, their management systems and execution. Also, it is aimed at developing scenarios for the near future, assessing the joker and «black swans». These uncommonly «artefacts» knits both with the human factor, and with the development of the world of technology, culture and the transformations of the natural world. The aerospace field also cannot develop without a clear understanding of its purpose and mission: ideological and moral support of life. The current crisis in the aerospace industry is due precisely to the fact that these pillars proved is much destroy. The fact is that humankind has lost an understanding of why the humankind needs space. As a result, there are numerous problems and accidents.

4 Discussion

Perception of the cosmos as another sphere of development, conquest of nature has led a person to a dead end. As K.E. Tsiolkovskiy noted, the Earth is the cradle of humankind. But man cannot always remain in the cradle. Space is the place and time of development not just «exploitation and development». Man is mastering himself, not space itself. And he can master it, only being a cosmic being. Only when man recognizes himself as a cosmic being involved in the development of the entire universe, he does acquire the moral right and meaning of entering into the «great cosmos». On the contrary, closing in on pragmatics, for example, trade in space territories, research conducted within the framework of corporate requests, which are only interested in financial benefits and social control, a person will not be able to advance into space. Moreover, he begins to suffer on the Earth. These are the place and time, in which he ceases to be a man-inhabitant of the Earth, a part of nature and culture, but becomes an automaton the part of the technologies. Attitude towards man as an impersonal automaton, so typical for the end of the XX and beginning of the XXI century, destroys the meaning of the development and existence of the aerospace industry. Especially negatively among the aerospace industry, the development of the industry is badly influenced by educational institutions that train inadequately competent specialistsAs a result, we have an uneven, crisis-evolving industry. In this industry, we have conquests and achievements alternate with breakdowns in accidents. Thus, the motivation of specialists in the aerospace industry, their professional and career development, is a general management of aerospace enterprises. However, it is far from perfect on the one hand, and far from understanding their purpose and mission on the other. Indeed, one cannot strive for anything unknown, and perfection is always relative. It also determines their final state and results, as well as the ways to achieve this state, the results. K.E. Tsiolkovskiy, other philosophers and cosmic scientists talked about this even more fully and definitely.

K.E. Tsiolkovskiy considered civilization as a single organism that undergoes several phases in the development. Each of phases is associated with the formation of new, more realistic and detailed ideas about happiness, the value of human life, the conditions of its implementation and meaning, as well as the transformation of man: immortalism (immortality), victory over time and autotrophy, victory over space, «interplanetarism» (interplanetary life). «For life in space, man and humankind need a new degree of development, a new-another-understanding of oneself and the world. The idea of the relationship between man and the cosmos was reflect in his cosmic philosophy: «The whole cosmos determines our life», he noted, «everything is continuous and everything is one ... It is hard to imagine that any part of it (the cosmos) sooner or later had no influence on us». «The universe would not have made sense if it had not been filled with an organic, sensible, sensible world [1]. Part of his work is his «cosmic ethics», which includes the development of ethical foundations of contacts with aliens, the recognition of the need for joint labor for the transformation of outer space. Exit to space expanses is an important aspect of the evolution of humankind, and the ideas of autotrophy, self-feeding of humankind, the connection of man and the universe, developed by K.E. Tsiolkovskiy, followed by V.I. Vernadskiy [2].

The emergence of man is an act of the greatest importance, a single during the geological history: he does not have anything analogous in the environment of the myriads of the preceding centuries [2]. At the same time, «The progress of organisms is continuous and therefore can not stop at a person», asserted K.E. Tsiolkovskiy; «Man is not» the crown of creation», In V.I. Vernadskiy thought, consciousness and life in their modern form must inevitably be followed by «super-consciousness» and «super-life» [2]. K.E. Tsiolkovskiy spoke about the future man, the «animal of the cosmos», directly assimilating in its nutrition the sun's rays and elementary substances of the environment and can be immortal. He and V.I. Vernadskiy stressed the importance of autotrophy «The consequences of such a phenomenon in the mechanism of the biosphere would be enormous. This would mean that a single whole - life - would again be divided, a third, independent branch would appear ... The human mind in this way would not only create a new great social achievement, but would introduce a new great geological achievement into the mechanism of the biosphere ... » [2]. In this case, K.E. Tsiolkovskiy takes life as a "circuit". He notes the helicity of life. This is for him the most common type of development. He distinguishes in it as stages of an ascending, more complicated development, as well as the stages of the subsequent "decomposition", "simplification", a return to a more elementary form. Further, after them follow the stages of the new revolution. There is an even greater complication, and then an even greater simplification, and so on - to infinity. K.E. Tsiolkovskiy develops the idea of N. Fedorov about the interconnectedness and interdependence of humankind's victories over space and over time. Immortality is possible only in outer space, infinite and inexhaustible in its energy and material resources. Only immortal creatures with an essentially transformed organism and the awareness of themselves and the world can survive in a variety of extraterrestrial environments, master and transform the Universe. Later, in V.A. Sukhovo-Kobylin described the active evolution of man, aimed at the cosmic expansion of humankind. This transformation presupposes «an infinite involution ... into oneself, deeper into» self-consciousness, up to such a «connection with oneself», such a spiritualization that this future organism becomes, as it were, spatial, supersensible («until the flesh disappears»), «Ethereal». This is very similar to the idea of the future «ray» humanity in the «theory of cosmic eras» by K. E. Tsiolkovskiy [3]. However, in their understanding, the cosmic nature of the phenomena of life and man, the demand for human activity in relation to the macrocosm, was combine with the natural rootedness of this life. So, V.I. Vernadskiy writes about the principle of F. Redi. «F. Redi claimed: every living organism comes from another living the same organism» [2]. He notes «Spontaneous generation that is, the genesis of a living organism due to inert matter, without the mediation of another living organism, is still seemingly logical for many scientists ... for a scientific explanation of life. With deep faith, beliefs are expressed that the direct synthesis of the organism from its material elements must be a necessary completion of the development of science. «However, in his opinion», of course, it is possible that they correspond to reality. You cannot consider them scientifically refuted. However, nothing indicates their likelihood. Nothing also indicates that the problem of spontaneous generation does not belong to the same series of searches as the problem of quadrature of the circle, the trisection of the angle, the «perpetuum» mobile, the philosophical stone». However, «The desire to resolve all these problems was not fruitless, it had very important consequences. It led to great new discoveries, but the most problems turned out to be unrealistic» [2]. K.E. Tsiolkovskiy considered civilization as a single organism, which takes several stages in development. In his opinion, humankind cannot «live forever in the cradle», and the output into the near-Earth space becomes the second (after the embryonic) phase of the evolution of humankind. This raises the question of the unity of the «terrestrial» and «extra-terrestrial» humankind (all «neo-humanity») and reproduction, the reproduction of man in space. «Space activity can be contrasted with the earthly and demand the formation of other ideas about happiness, the value of human life, the conditions of its implementation and meaning» [3]. The third phase is the resettlement of people throughout the Galaxy, where in time people will become truly «cosmic animals», that is, they can live without special means of protection in an open space: they need a new degree of development, and peace [4, 5]. On the way to this understanding, to a new awareness of yourself and the world. In the opinion of K.E. Tsiolkovskiy, «The coming thousands and millions of years will improve the nature of man and his social organization. Humankind will turn as if into one powerful being» [3]. «There will come an unification, they will cease due to this war, since there will be no one to fight with. The happy social order, prompted by geniuses, will force technology and science to go forward with unimaginable speed and with the same speed to improve human life. ... There will be full scope for the development of both public and individual properties of a person that does not harm people», K.E. Tsiolkovskiy wrote [3]. The ethics of the conscious beings of the cosmos, according to K.E. Tsiolkovskiy, «consists in that there is nowhere any suffering: neither for the perfect, nor for other immature, or beginnings animals» [5]. He noted the importance of «space selection», wrote that it is necessary to strive for a painless repayment of the life of unsuccessful genetically humans and animals, to improve them by encouraging the birth of more perfect and to charity towards everything imperfect, to animals, and humans [6]. He also believed that all particles of the universe have infinite complexity: the universe has developed or complicated for an infinite time, that is why for the person there are and can be accessed the most diverse materials of different infinitely removed from today, including «parallel existing» worlds and epochs. This will lead to the fact that a person realizes not only his multiplicity and infinity «in itself», but also his own dialogue and unity with the world. L.M. Gindilis noted that an important part of the concept of K.E. Tsiolkovskiy «is the notion that civilizations do not develop in isolation, but in close cooperation, to which new members, new communities of sentient beings join as they develop. This public organization is built on a hierarchical principle. «Unions of the nearest suns are formed, unions of unions, etc. Where the limit to these alliances is difficult to say, since the universe is infinite» [7]. He also did not exclude the possibility of hidden contact phenomena, visits and interference of highly developed civilizations in earthly affairs. «Sometimes we see unusual phenomena», he wrote. «Are they the result of the activity of surviving sentient beings of other epochs» and other more developed and perfect civilizations?» [7]. All this suggests that K.E. Tsiolkovskiy and Russian cosmism as a phenomenon was a philosophy of action and faith: faith in God as a belief in Life was backed up by scientific research and data that would not be «deciphered» at first, and probably, not even in this century. However, the enormity of the problems raised and solved by the Russian cosmos builds the way for other sciences and research that do not exist nowadays, which modern humankind sometimes does not even guess.

5 Conclusions

Thus, aerospace enterprises in the modern world badly need to conduct a serious "foresight-audit", an external and internal audit of their being and development. Such an audit is necessary at the level of values, goals and mission of their life. Such an audit was also need at the level of the management system of aerospace enterprises and groups of specialists and individual specialists. External and internal audit is necessary for that. To assess the static dynamic aspects of aerospace enterprises, including the static and dynamic capabilities of enterprise managers. It is also very important to assess the prospects and limitations of development at the level of each individual specialist. Time and place, when the aerospace enterprises could afford a "generalized", impersonal and morally pointless approach, do not exist anymore. The cosmos will be "mastered" only with the mastering of oneself by man, only when harmony of man and the world is reached [8].

References

1. K. E. Tsiolkovskiy, *Mechty o Zemle i nebe* [In Russian] (Tula: Priokskoe knijnoe izdatelstvo, 1986)

2. A. V. Sukhovo-Kobylin, Filosofiya dukha ili sotsiologiya. *Russkiy kosmizm: antologiya filosofskoy mysli* (Moscow: Pedagogika-Press, 1993)

3. A. A. Medenkov, M. A. Milovanova, N. L. Fetisova, Filosofskiye problemy mezhplanetnykh poletov. *Idei K.E. Tsiolkovskogo: proshloye, nastoyashcheye, budushcheye. XLVII Nauchnyye chteniya* (Kaluga: K.E. Tsiolkovskiy CDSH, 2012)

4. O. Gazenko, Kosmonavt dolzhen ostavat'sya chelovekom Zemli [In Russian], *Science and Life* **4**, 24 (2006)

5. K. E. Tsiolkovskiy, *Ocherki o Vselennoy* [In Russian] (Kaluga: The Golden Alley, 2001)

6. K. E. Tsiolkovskiy, *Kosmicheskaya filosofiya* [In Russian] (Moscow: URSS, 2001)

7. K. E. Tsiolkovskiy, Kosmicheskaya filosofiya. *Archive of the USSR* [In Russian] (Moscow: Academy of Sciences, 1935)

8. M. R. Arpentieva, Ksenopsikhoterapiya [In Russian], *Philosophy & Cosmology* **1(15)**, 163 (2016)

Time-saving method of orbital thermal regime calculations of nanosatellites as exemplified by a 3U CubeSat

Vasily Gorev[1,*], *Vitaly* Prokopyev[1], *Yury* Prokopyev[1], and *Alexey* Sidorchuk[1,2]

[1]Novosibirsk State University, 630090, Pirogova str., 2, Novosibirsk, Russia
[2]OKB Fifth Generation Ltd., 630090, Nikolaeva str., 11, Novosibirsk, Russia

Abstract. A time-saving approach to perform technical calculations of thermal conditions of orbital motion of 3U CubeSat nanosatellite was applied, which made it possible to make the thermal calculations of a satellite with simple structure geometry using MatLab and SolidWorks Simulation. Passive thermal regulation facilities are sufficient for a 3U CubeSat to provide thermal conductivity of the case's structural elements and to remove heat from the lighted surface and internal components to the satellite's shadowed surface. Application of spectrally selective coatings allows narrowing the range of surface temperatures of 3U CubeSat.

1 Introduction

The design of the thermal operating conditions is one of the main tasks in satellite development procedure. For the regular operation of the satellite equipment, it is necessary to maintain the thermal conditions of the satellite functional elements in a certain narrow operational range against the extreme conditions realized in space. The initial requirements for the design of the satellite's thermal regime are the conditions of its operation in space, its power-to-weight ratio, and the operating modes of the heat producing parts [1].

All space equipment requires a precise development because of impossibility of any kind of reparations after launch. Therefore, comprehensive ground-based tests of onboard systems as well as structural elements of satellites, computational techniques come into importance. They make it possible to optimize many parameters of the device on early stages of design. Calculating the thermal conditions of the spacecraft is a standard and required task; there is a range of instruments for performing such calculations. There are also specialized software packages for calculations of radiative heat transfer, such as Thermal Desktop, ESARAD / ESATAN TMS [2] and other software products. In general, in the family of software products for scientific and technical calculations, there is a tendency splice it with MCAD development environments, thus creating a unified MCAD development environment with numerical studies facilities. This fact is caused by desire to automate the mostly manual transfer of product geometry from a development software to a calculation program.

* Corresponding author: vasily.gorev@gmail.com

This paper discusses the problem of developing a time-saving method for in-orbit thermal conditions calculation of nanosatellite. It is proposed to stay within capabilities of MCAD environment where the spacecraft is being developed, i.e., SolidWorks Simulation. This approach makes it possible to reduce significantly the time for transferring and adapting the satellite model geometry to calculation program. Nevertheless, it is still sufficient for structure development to obtain a necessary information about thermal conditions of nanosatellite.

2 Calculation procedure

2.1 Calculation of radiation absorbed by a satellite during orbital motion

The procedure for calculating the nanosatellite thermal conditions consists of two parts:
1) calculating the power absorbed by the faces of a satellite performed in the MatLab environment;
2) calculating the heat transfer under structural elements of the case of a satellite and the heat dissipation in the outer space realized using the SolidWorks MCAD tool – SolidWorks Simulation.

During the first step of the procedure, when modeling the satellite radiation conditions, a circular Low Earth Orbit was considered; the main external parameters of the task are the altitude H of the orbit and the angle B between the satellite's orbit plane and the direction to the Sun [1, 3]. For this calculation, H is assumed 450 km that approximately corresponds to the International Space Station orbit.

Let us consider the orientation where the satellite will be directed by a small face of the 1U case toward the Earth, and one of the large lateral 3U faces will be perpendicular to the orbital velocity (Figure 1, a). To calculate a satellite insolation regime, all elements protruding from the case, such as antennas and folding panels of solar batteries, are ignored. The intensity of absorbed power is calculated on the satellite case outer faces for various materials: anodized aluminum alloy, photovoltaic converters, and the face panels' surface covered with the spectral-selective paint.

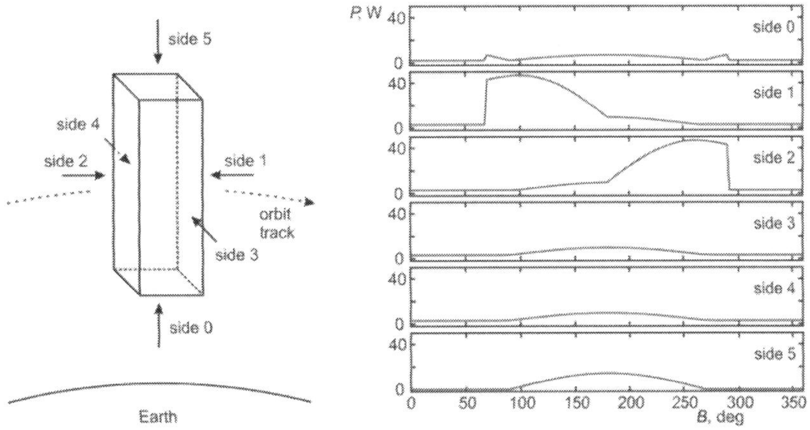

Fig. 1. Earth-pointing and relative motion (left) orientation of the 3U structure. Total incident power radiation (right) applied to satellite surfaces, depending on the position in orbit at $B = 0°$.

The calculation takes into account three radiation sources: the solar radiation, the Earth's albedo, and the Earth IR radiation. The solar radiation is calculated as the radiation of a black body with a temperature of 5762 K, which approximates the solar radiation spectrum with the sufficiently high accuracy [4]. The solar radiation average intensity is assumed to be 1367 W/m^2 [1]. To calculate the magnitude of the radiation reflected by the Earth, the approximate calculation method recommended in work [4] was applied. This method conforms the experimental data obtained within the framework of the Earth Radiation Budget Experiment (ERBE) project [1, 5]. The value of the Earth infrared radiation is assumed to be 245 W/m^2 at the altitude of 30 km above the Earth's surface [1, 5], and it is recalculated for the device orbit altitude.

2.2 CubeSat calculation models

To calculate the thermodynamics of the nanosatellite, several CAD models of the CubeSat with various simplifications regarding the actual design of the 3U satellite available at this design stage was developed. Thermal conditions were calculated for models of different design simplification degree (from simple to complex).

The first simplified 3U-type model (*model-1*) is a single body combining frames and face panels made of anodized 2024 aluminum alloy; the absorption coefficients of solar radiation and the IR emission are taken from [6, 7]. On the outer surface of the model, photovoltaic modules are located, as the isolated parts of the surface. Inside the model, there are no electronic components and boards, only structural elements for their fastening (Figure 2, a). This approach simulates the design of the spacecraft in which aluminum panels coated with a thin layer of the copper-plated FR-4 are used - a common material for PCB, with high heat dissipation (HA80 and other similar materials), applied, for example, in LED lamps. In terms of the thermal conditions of the spacecraft and photovoltaic modules, using the FR-4 or aluminum-based panels can have a qualitatively different effect. In the case of aluminum panels, a high thermally conductive case minimizes the instantaneous case temperature difference; it conducts the heat from the illuminated sides to the unlit ones, and thus it facilitates the effective heat dissipation in space.

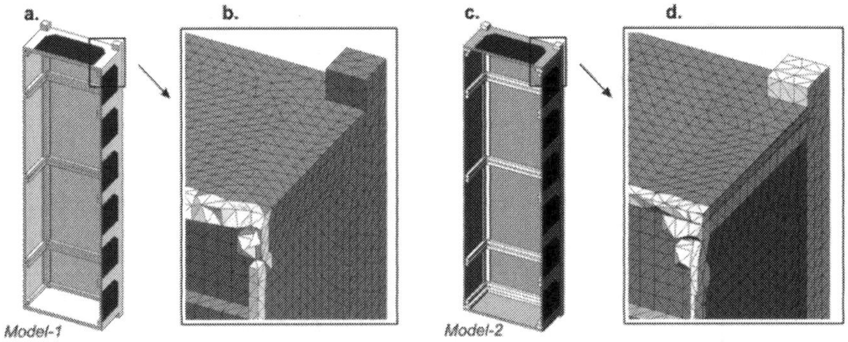

Fig. 2. CubeSat 3U case models. A one-piece case made of aluminum alloy (a) and its finite element mesh (b), a case made of copper-plated FR-4 panels on an aluminum alloy frame (c) and its finite element mesh (d).

On the other hand, the FR-4 can be much more convenient when installing the electronic components on the solar panel to control the inclusion of batteries in the satellite power supply system. Further, on *model-2* with the side panels made of copper-plated FR-4 (Figure 2, c) was considered. The panel consists of three layers: a fiberglass composite layer 1.6 mm thick is coated on both sides with layers of 0.2 mm copper. In the real construction, the FR-4 can have two, four or more layers of copper of smaller thickness. The panels' thermodynamics at the typical time of this heat exchange process is two-dimensional with good accuracy, so only the total thickness of the copper layer and the fiberglass composite layer is important. The calculated grid on thin planar elements has one cell in thickness direction, i.e., it is two-dimensional (Figure 2, d). The thermal resistance of the contacts between the details is not taken into account while calculating.

3 Results and discussions

The CubeSat dynamic calculation of the thermal conditions of *model-1* was made for the orbits corresponding to the values of the angle $B = 0^{\circ}$, 45°, 70° (Figure 3). The B angle = 70° corresponds to the point of the maximum average-orbital temperature for the H orbit altitude = 450 km for the studied satellite orientation in space, because it is the smallest one among the B angles, at which the satellite does not enter the Earth shadow. As the initial conditions for dynamic calculations, one uses the solution of the stationary problem for the equilibrium case at the initial point of the path - in the middle of the orbital eclipse period.

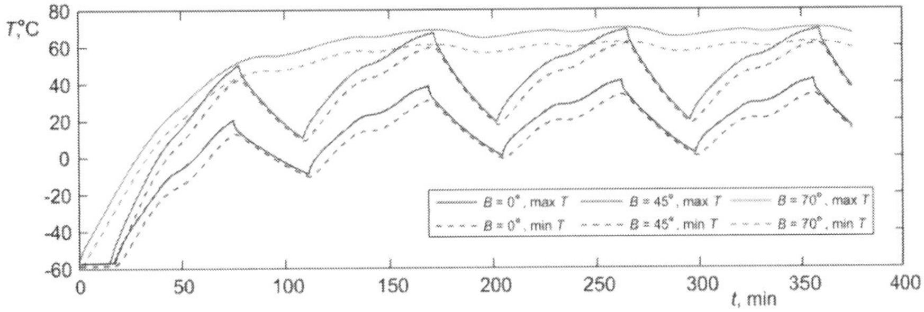

Fig. 3. Graphs of the minimum and maximum satellite surface temperatures for orbits with $B = 0^{\circ}$, 45°, 70°.

Figure 3 shows the graphs of the minimum and maximum temperatures of the outer surface of the satellite model's case. The initial part of the graph corresponds to the transient process from the equilibrium case for the orbital eclipse period of the initial conditions to the satellite's periodic thermal conditions. In the steady-state periodic part of the graph one can see the features of the thermal conditions associated with the shape and orientation of the satellite (they are clearly seen in the graphs for $B = 0^{\circ}$); in the areas of the spacecraft temperature growth, observing a bend corresponding to a decrease in insolation associated with the orientation of the satellite by a small face on the Sun. The temperature fluctuations in the case of the B angle B 70° are not big, because the spacecraft on this orbit doesn't enter the shadow of the Earth, and they are caused by a change in the orientation of the spacecraft relative to the Sun and by the periodic illumination from the Earth's albedo. One should note that the difference between the minimum and maximum instantaneous temperature of *model-1* does not exceed 10°C, which is achieved due to the high thermal conductivity of the satellite's case.

For *model-2* of the spacecraft, two versions of the thermal calculations for the steady-state orbital case concerning the satellite thermodynamics for $B = 90^{\circ}$ was made (Figure 4): 1). The side panels are made of 2 mm fiberglass composite; there is no copper deposition (Figure 4, b). In this case, the heat removal from the illuminated panel to the rest panels is difficult due to the low thermal conductivity of the fiberglass composite, observing some areas with a temperature exceeding 200°C on the illuminated surface. The temperature maxima are located between the aluminum fins of the frame, to which the panel adjoins. The absorption coefficient of the sunlight range and the emission factor of the IR radiation of the panel surfaces are assumed to be equal to the parameters of the 2024 anodized aluminum alloy (shown in Fig. 4. 2). The side panels are made of copper-plated FR-4, as described above. Two copper layers with a total thickness of 0.4 mm on the side panel reduce the maximum temperature of the illuminated side of the satellite to 60°C (Figure 4, c), which is close to the case of *model-1* for the same orbit (Figure 4, a). Figure 4, d shows the surface temperature field of the CubeSat model for the case identical to the one shown in Figure 4, c, except for the optical properties of the surface of the side panels.

Assuming the coefficient of the sunlight absorption and the IR radiation emission factor to be equal to the parameters of the spectral-selective paint (Figure 4, d). Covering the panels with the selective paint in the intervals between photovoltaic modules makes it is possible to decrease the upper limit of the temperature range of the satellite's case up to 25°C, the lower one to 10°C.

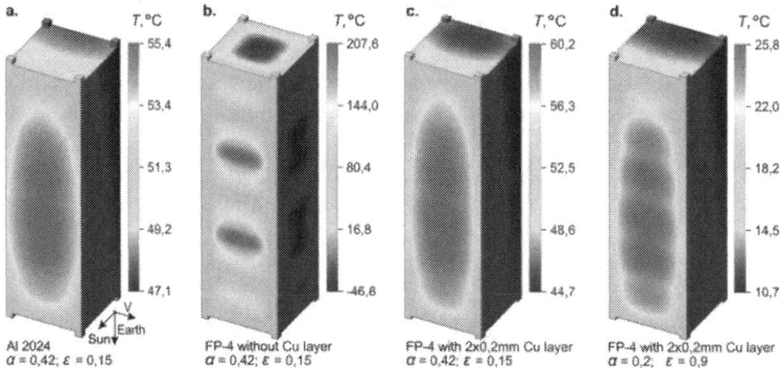

Fig. 4. Spacecraft model surface temperature fields. A solid case made of aluminum alloy (a), a case made of FR-4 panels without copper-plating (b), with copper-plating on an aluminum alloy frame (c). Case (c), covered with spectral-selective paint (d).

4 Conclusion

The time-saving method of technical calculations of thermal conditions is described as exemplified by 3U CubeSat nanosatellite with simplified structure geometry. It consists of two procedures: calculating the boundary conditions for the surface elements in the MatLab software package and calculating the thermal conductivity of the satellite structure and the IR radiation of the satellite's surface via SolidWorks Simulation.

The characteristic time of radiation thermal stabilization for 3U CubeSat is comparable with the period at Low Earth Orbit. The time of satellite surface temperature equilibrium setting is much less then orbital period for the case of 2 mm thick aluminum alloy face panels or FR-4 face panels with 0.4 mm total thickness of copper layer. Therefore, passive facilities are sufficient for 3U CubeSat thermoregulation.

Application of spectrally selective coatings allows narrowing the range of surface temperatures of 3U CubeSat. Coating with $\alpha = 0.2$, $\varepsilon = 0.9$, instead of the anodized 2024 alloy ($\alpha = 0.43$, $\varepsilon = 0.15$), reduces the maximum surface temperature of the satellite from 60.2 °C to 25.8 °C.

This work was supported by the Ministry of Education and Science of the Russian Federation: Project No. 14.575.21.0154, Project Identification No. RFMEFI57517X0154.

References

1. *Spacecraft thermal control handbook. Volume I: Fundamental technologies* (Edited by D. G. Gilmore, The Aerospace Press, 2002)
2. K. Yang, TFAWS, Pasadena, CA, (2012)

3. *Space Mission Engineering: The New SMAD* (Edited by J. R. Wertz, Microcosm Press, 2011)
4. R. E. Smith, G. S. West, NASA Technical Memorandum 82478 (1983)
5. B. R. Barkstrom, E. F. Harrison, G. L. Smith, R. D. Cess, JASR **9**, 7, 75 (1989)
6. J. H. Henninger, NASA Reference Publication 1121 (1984)
7. J. L. Golden, *NASA Conference Publication* 3257, 61 (1992)

Thermal deformation of 3U CubeSat in low Earth orbit

Vasily Gorev[1,*], *Anatoly* Pelemeshko[1], *Alexander* Zadorozhny[1], and *Aleksey* Sidorchuk[1,2]

[1]Novosibirsk State University, 630090, Pirogova str., 2, Novosibirsk, Russia
[2]OKB Fifth Generation Ltd., 630090, Nikolaeva str., 11, Novosibirsk, Russia

Abstract. The impact of uneven heating on a satellite structure in low Earth orbit has been considered using the example of 3U CubeSat. The calculations showed that the thermal deformation of CubeSat structure in orbit caused a deviation between normals to opposite small satellite sides of about 0.03°. Such a deviation is commensurate with the required satellite pointing accuracy approximately 0.1° necessary for satellite laser communication. It means that to solve similar problems in the CubeSat designing that require such or better CubeSat pointing accuracy, it is necessary to take into account the expected satellite structure thermal deformation.

1 Introduction

Novosibirsk State University develops a universal modular platform for nanosatellites compatible with CubeSat. The platform should also support optical (laser) communication. The creation of the positioning system module is one of the most important tasks of the project, since it provides the satellite with pointing accuracy required to operate an optical (laser) communication system. The satellite's pointing accuracy with a rigid laser fixing on its structure should not be worse than 0.1° to direct the laser beam to the ground receiving station [1].

The deformation of the satellite structure due to temperature variations on its surface during the orbital motion is one of the factors affecting pointing accuracy.

While operating in orbit, CubeSat absorbs radiant energy originating from external sources, and emits thermal radiation from its surface into the surrounding space. Uneven radiation heating leads to uneven satellite heating, which can vary widely depending on its design.

In developing a high-precision pointing system it is necessary to understand the impact of heat fluxes on the satellite's surface while in the Earth's orbit and corresponding deformation on the satellite structure in order to make constructive decisions to minimize or compensate for them. The aim of this article is to study the impact of uneven satellite heating in low Earth orbit (LEO) on structure deformation according to the example set by 3U CubeSat.

*Corresponding author: vasily.gorev@gmail.com

2 CubeSat pointing and its design

The positioning system of a modern CubeSat generally contains multiple sensors such as Sun sensors, Earth sensors, horizon sensors and star trackers (Figure 1, a). This article describes the configuration of a satellite directed by the 1U side to face the Earth's surface. The choice of such a positioning can be convenient for an optical device with a large aperture having an extended dimension in the direction of the observation vector. The laser communication optical receiver may also require the satellite assembly stretched along the Earth's radial direction.

The positioning sensors and laser communication unit (or observation unit) can be fixed to either of opposite small satellite sides.

This means that, due to thermal deformation, there will be an additional angle between the optical axis of the laser communication system (or observation) and the optical axis of the star tracker, as the most accurate of the positioning sensors, which was missed at calibration. Thus, the positioning system will have a systematic error.

For ease of calculation, it is assumed that the orbital motion vector is perpendicular to one of the 3U sides.

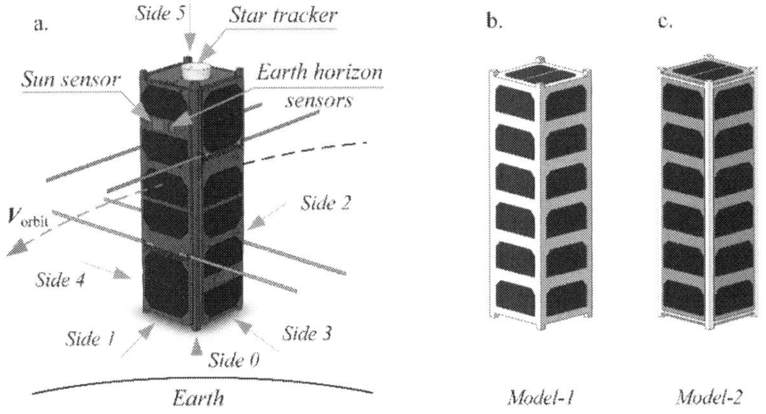

Fig. 1. 3U CubeSat positioning relative to the Earth and the positioning system sensors' configuration on the satellite surface (a). CubeSat structure design (b, c).

This article describes two versions of the 3U CubeSat structure (Figure 1, b, c). The first simplified model-1 is a solid structure combining the framework and panels made of anodized aluminum alloy 2024. For this alloy, the solar absorption and IR radiation emission coefficients ($\alpha = 0.42$ and $\varepsilon = 0{,}15$) were taken from study [2]. In model-2, the satellite's side panels consist of a 1.6 mm thick FP-4 glass epoxy, which are coated with a 0.2 mm copper layer on both sides. In order to compare only the impact of thermal conductivity, the α and ε coefficients for the model-2 made of FR-4 are equal to the coefficients of the model-1 made of the alloy 2024.

In each model under study, the photoelectric transducers on the sides of the satellite structure are selected areas on the surface, which are also taken into account in the calculation with their own coefficients $\alpha = 0.92$ and $\varepsilon = 0.9$. There are no electronic components and printed circuit boards inside the models.

3 Thermal regime calculation procedure

The thermal regime calculation procedure for the satellite is performed using the SolidWorks Simulation CAD tool taking the values of the absorbed radiation power calculated in the MatLab environment as boundary conditions.

Modeling of the CubeSat radiation regime was carried out for circular low Earth orbit. The orbit height H and the angle between the plane of the satellite's orbit and its direction to the Sun B are the main external parameters of the task [3, 4]. For this calculation, the orbit height H was assumed to be 450 km, which roughly corresponds to the orbit of the international space station. In order to cover the possible range of angles B = 0°-90°, a family of orbits perpendicular to the ecliptic plane is considered.

The following radiation sources are included in the calculation: solar radiation, Earth albedo and Earth infrared radiation.

Depending on angle B, different thermal regimes of the satellite orbital motion can be observed. Thus, for a satellite with an absolutely black surface, the mean orbital radiation power and mean orbital temperature on its surface are calculated. The satellite is believed to come into thermal equilibrium instantly (Figure 2).

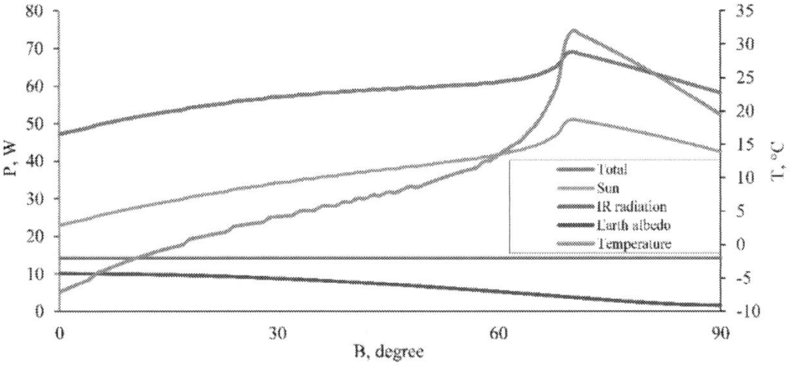

Fig. 2. The equilibrium mean orbital temperature of the satellite with an absolutely black surface (green line) and mean orbital power of the heat flux on different components on the CubeSat surface.

The range of *B* angles can be divided into two parts. In the subrange B = 0° – 70°, the satellite enters the Earth's shadow and, therefore, experiences considerable temperature fluctuations. In the subrange B = 70° – 90°, during the entire orbital flight the satellite is illuminated by the Sun, and the mean temperature does not vary widely. The subranges boundary for mean orbital power and temperature is shown on the figure in the form of a tooth at point B = 70°.

It can be seen that with a given pointing the satellite gets the greatest amount of energy in the orbit corresponding to the angle B = 70°. When angle B increases, the amount of energy reaching the satellite surface is reduced. This is due to the satellite pointing and to the change in the Earth's albedo, because if B is about 90°, the proportion of the Earth albedo in the thermal balance of the satellite is much less than the other components.

Table 1. Distribution of the absorbed power on the 3U CubeSat sides.

Side №	**0**	**1**	**2**	**3**	**4**	**5**	**Σ**
P_{fv}, W/м^2	199.7	93	93	1399.3	90.4	0	-
P_{A2024}, W/м^2	34.9	16.2	16.2	612.6	15.1	0	-
S_{fv}, м^2 *10^{-3}	5.45	16.6	16.6	16.6	16.6	5.45	77.3
S_{A2024}, м^2*10^{-3}	4.55	16.37	16.37	16.37	16.37	4.55	74.6
P_{total}, W	**1.25**	**1.81**	**1.81**	**33.26**	**1.75**	**0**	**39.9**

However, satellite mean orbital temperature in the orbit B = 70° and in the orbit B = 90° near the terminator does not differ qualitatively and quantitatively. Therefore, to simplify the calculation of thermal deformation, the thermal regime in the orbit at B = 90° is considered, which roughly corresponds to the solar-synchronous orbit passing over the terminator line. From the point of view of the satellite thermal regime and, accordingly, for thermal deformation, this formulation of the problem is stationary. In the presented satellite orientation, the main energy flux falls only on one CubeSat side.

The absorbed radiation power surface density for CubeSat surface materials ($\boldsymbol{P_{fv}}$ for photoelectric converters, $\boldsymbol{P_{A2024}}$ for anodized aluminum alloy 2024) and the total radiation power absorbed by the satellite sides are presented in Table 1

As was supposed, the absorbed radiation power has a maximum value on the side, which faces the Sun.

4 Evaluation of satellite structure thermal deformation

Static analysis of the CubeSat thermal deformation was performed by SolidWorks Simulation software for two models with varying degrees of detail described above.

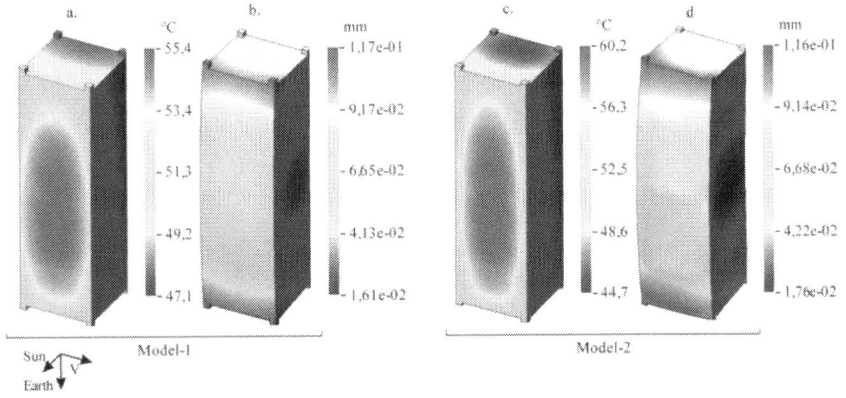

Fig. 3. Basic mode: a, c – surface temperature field, b, d – framework deformation field.

To calculate the displacement field of the structure grid caused by thermal deformation (Figure 3), the results of the thermal calculation for the orbit B = 90° were used. The surface temperature difference dT = 8,3 °C and 15,5 °C for the satellite with solid aluminum case and for the satellite made of FR-4, respectively, is explained by the different thermal conductivity of the model-1 and model-2 panels.

It can be seen that specified points on the satellite structure are displaced from their equilibrium positions by more than 0.1 mm. Herewith, the maximum displacements of the model-1 and model-2 points are not qualitatively different, since FR-4 fiberglass composite has a coefficient of thermal deformation comparable with metals.

The satellite construction has a pronounced bend in one direction, because one of its 3U sides is heated more than others. In addition, the problem has symmetry with respect to the Sun-Earth-satellite plane. In this case it can be asserted that the 1U sides have much less relative deformation than the heated 3U side, therefore we will consider them approximately flat.

To estimate the angle γ between normals to 1U sides, 4 corner points are selected on the small sides of the layout (Figure 4). For each point, their initial and displaced coordinates in the deformed state were determined. The selected points form four triplets along which it is possible to construct four pairs of vectors directed from each corner of the 1U side along the 3U side. For each pair of vectors, a normal was constructed by its cross product. Thus, for the opposite small sides, four normal vectors are constructed, and the angles between the corresponding normals of opposite sides are determined.

To solve the pointing problem in which the laser module and the star tracker are fixed on the satellite structure without any compensating technical solutions, the obtained deviation angles will increase the satellite attitude system error. Angular deviation (0.03°) makes a significant part of the required pointing angle accuracy to the receiving station (0.1°).

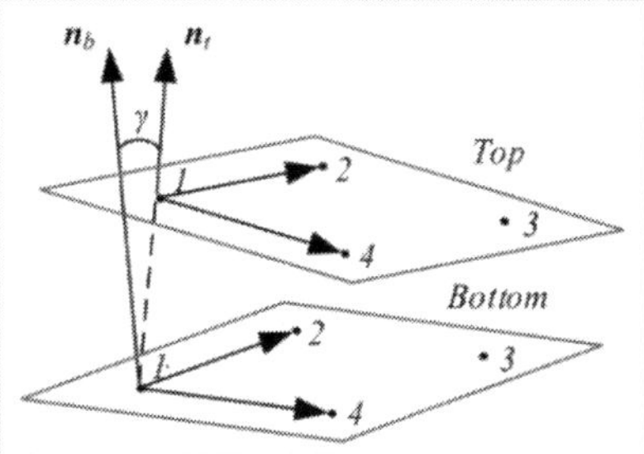

Model-1					
Point	*1*	*2*	*3*	*4*	***Mean***
α	0,024°	0,028°	0,028°	0,024°	**0,026°**
Model-2					
Point	*1*	*2*	*3*	*4*	***Mean***
α	0,030°	0,035°	0,035°	0,030°	**0,033°**

Fig. 4. Scheme for calculating the angle between the normals to the 1U sides. Angular deviation γ between the corresponding normal vectors.

Displacement of the specified points due to thermal deformation and the angle between the normals to the 1U surfaces depends on the difference between the maximum and minimum temperatures over the satellite surface (dT). As the thermal conductivity dT of the structure will decrease, accordingly, the displacement values and the angles between the normals will decrease too.

In designing a nanosatellite, it is necessary to take into account that satellite components thermal deformation can affect the satellite positioning system. A thermal deformation impact on optical system can be reduced by applying a number of technical solutions that make it possible to ensure the constant angles between the optical axes of the optical system units, in particular by the increase in the thermal conductivity of the satellite structure.

5 Conclusion

The impact of uneven satellite heating on the satellite structure deformation is analyzed by the 3U CubeSat in LEO example. It is shown that the thermal deformation of the CubeSat structure, which is in orbit of 450 km high and angle B = 90°, causes a deviation of about 0.03° from the normals to the opposite small sides of the satellite. Such a deviation is commensurate with the required satellite pointing accuracy on the order of 0.1° necessary for laser communication. Thus, in CubeSat designing to solve problems requiring the stated or better satellite pointing accuracy, it is necessary to take into account the expected satellite structure thermal deformation.

References

1. S. W. Janson, R. P. Welle, T. S. Rose, D. W. Rowen, D. A. Hinkley, B. S. Hardy, S. D. La Lumondiere, G. A. Maul, N. I. Werner AIAA/USU Conf. SSC15-III-1 (2015)
2. J. L. Golden, NASA Conference Publication 3257, 61 (1992)
3. *Space Mission Engineering: The New SMAD* Edited by J. R Wertz, D. F. Everett, J. J. Puschell, Microcosm Press, (2011)
4. *Spacecraft thermal control handbook. Volume I: Fundamental technologies* (Edited by D. G. Gilmore, The Aerospace Press, 2002)

MATEC Web of Conferences **158**, 01014 (2018) https://doi.org/10.1051/matecconf/201815801014
VI International Forum for Young Scientists "Space Engineering 2018"

Can MOOC help to prepare a well-trained specialist for aerospace industry?

Veronica Ivanova*, *Kseniya* Mertins, *Yulia* Barabanova, and *Irina* Yurkina

National Research Tomsk Polytechnic University, 634050 Tomsk, Russia

Abstract. Digitalization of education is the main trend of modern society. The ideology of education "lifelong learning" invests the opportunity to improve skills and develop competencies without leaving home. The article discusses the possibility of using MOOC for training specialists in the aerospace industry. The analysis of the MOOC and specialized courses developed and followed in the TPU, as well as analyzed the education of students is done.

1 Introduction

Mass-open online courses (MOOCs) are named among the 30 most promising trends in the development of education until 2028. It is believed that they open up new opportunities in the field of distance education increasing the relevance of the topic of quality online courses [1]. The MOOC concept is based on the key principles of the new theory of learning – connectivity: diversity of models, approach to learning as a process of network formation and decision-making, learning and cognition as a dynamic process.

Features of MOOC:

• training is free of charge or conditionally paid (for the certificate) and voluntary;

• a large number of students on the course (hundreds and thousands);

• emphasis on independent work, self-control and mutual control;

• openness-MOOC is accessible through the Internet 24x7x365 without any restrictions;

• multimedia-MOOC use audio, video (including interactive), 3D-worlds and many other modern technologies.

Presentation of training information:

• theoretical material is represented by lectures (in the recording), which are divided into parts with a duration of 2 to 15 minutes. Lectures end with a test, quiz or control questions for checking understanding;

• workshops in the form of various tasks: solving mathematical problems, essays, discussions, creative projects, team projects, work in virtual laboratories, etc.

Organization of control:

• focus on independent work. Independent and cross (mutual) evaluation of each other's work is used to monitor the activity of students,

• the total final control measure for the certificate [2].

* Corresponding author: kurator@tpu.ru

2 Nowadays situation

Major American and British universities actively participate in the creation and dissemination of educational resources, including open online training courses, and are developing the market of electronic educational services. [3]

Some of the researches are sure that MOOCs provide free access to high-quality learning materials, offered by elite universities. They are conceptualized as online learning environments in which participants worldwide can create, research, and share open educational resources. However, the opponents consider them as a passing trend that might damage higher education, including research and accreditation.

Another debate relates to the way students learn and whether MOOCs can facilitate deep and meaningful learning. Mackness, Mak and Williams (2010) asserted that the more autonomous, diverse and open the MOOC is, the more the potential for students' learning to be limited by the lack of structure, support, and moderation which are normally associated with a regular course. [4]

One more series problem is lack of mechanisms for assessing the quality of e-learning and the MOOC by consumers.[5]

The analysis of the MOOCs, covering the issues of training specialists for the aerospace industry, showed the following. On two Russian platforms (Stepik and Universarium) there are five courses related to this field, on the foreign platforms 8 courses were detected.

3 TPU and MOOC

TPU has a positive experience of MOOC using. Since 2016, at TPU's own platform http://edx.tpu.ru students are trained with the MOOC using. MOOCs are used within the framework of the disciplines "Professional English Training" and "Foreign Language". Students study "Introduction to petroleum engineering", "Myths and Facts About Rocks» [6]. 85 percent of TPU students who train in this discipline received a certificate for the successful completion of the course.

Teachers use MOOC as an additional material for the discipline, applying different technologies:

• technology "inverted class". Independent work of the student involves studying the video lectures of the course and performing evaluation activities. During the classroom work complex aspects of the course are worked out and practical assignments / laboratory works are performed;

• the organization of independent work only on the basis of the MOOC;

• formation of professional communication skills (English language) on the basis of the MOOC. A glossary (thematic dictionary) of the basic terms of the course in English is formed and developed. The work in the classroom is aimed at fixing professional terms in English, and in the course of independent work, the students carry out the tasks of the course.

Summative assessment of the discipline is organized at the discretion of the teacher:

1. offset on the discipline on the basis of the final result at the rate;
2. the final result at the rate is taken into account as a written part of the offset, and the oral part of the test involves the protection of the project in English within the framework of the course.

This experience can be passed on improving the quality of future engineers training, including the field of space engineering. The complexity of the material of engineering disciplines does not allow to limit examination by tests. Special types of tasks oriented on practicing practical skills required. The system of TPU MOOCs tasks contains tasks for mutual verification, as well as practice-oriented tasks with automatic verification (including the checking of entrepreneurial competencies).

Here is an example of the course "Logistics for yourself. To develop entrepreneurial thinking is a popular science course, but also it helps to obtain specific skills (through a system of tasks / business cases). It lasts for 5 weeks and contains four practical calculation tasks, one laboratory work and analysis of four thematic business cases.

Task № 1 Determination of the required number of days of additional insurance reserve for the enterprise. In Task № 2 Student need to choose a store in which it is more profitable to buy products for the preparation of a particular dish, and to calculate the minimum cost of this dish. Task № 3 devoted to determination of more profitable warehouse: own or hired, and to calculation of the cost of the selected type of warehouse and the point of indifference of cargo turnover. Task № 4 poses to calculate and form a template in the program MS Excel, which can then be used to make decisions to optimize the real business processes of the company.

- virtual laboratory work "Choice of warehouse ownership" is implemented in the format of interactive video – contains elements of active influence of the viewer on the reproduced video and allows to organize a non-linear scheme of viewing video to simulate live communication of the teacher and the student, the organization of an interactive dialogue between the viewer and the lecturer.
- the average percentage of people who have successfully completed MOOC depends on a number of factors:
- on the direction to which the course belongs (art, Humanities, business and management, computer technology, biological Sciences, psychology, physics, mathematics and logic, engineering Sciences, etc.),
- on the support of the course by teachers/curators (frequency of answers to questions, the presence/absence of a thematic community within the course, the atmosphere in the course, etc.),
- on the purpose of the course (education, educational – in such courses % of successfully completed training will be less than in courses that are aimed at practicing some practical aspects (PC courses, retraining, etc.),
- career guidance (bonuses for successful completion of the course, for example, extra balls on admission, self-determination and ect.).

As a rule, students already come with a request for certain points (practicing skills, obtaining a certificate, bonuses, their self-determination, etc.). They are already motivated to pass, because it will either "pump them", or they will receive some bonuses for themselves.

Here is a comparison of two most wide used on-line technologies MOOC and Moodle (Table 1.)

Table 1. Distinctive features of MOOC and Moodle.

Parameter	MOOC	Moodle
Course name and audience focus	Original and attractive name	As a part of the educational program, as a rule, coincide with the modules or parts of the discipline
Quantity of students	Massive participation (more then 1000, otherwise the will not be opened)	From 1 person, the course holder decides (in accordance with the curriculum)
Structure of the course	Information block Training and Practical Unit The control unit Organizational unit	Availability of large quantity of material Temporary access to materials after compulsory course units

	Communication block Dosed temporary access to materials after compulsory course units	(subject to curriculum restrictions)
Verification materials Interactive assignments	Ability to assess competencies Inclusion of simulators for assessing professional activity	Problem-oriented learning, gaming, social networks, web 2.0, virtual laboratories, the creation of educational content (teacher + student), peer-to-peer peer evaluation mechanisms, etc.
Period of passage	3-6 weeks	From one week till one semester
Time resources for development	Till 100 hours per week	The ability to reduce time costs by copying elements
Individuality and competences of the teacher	Demand for charisma, professionalism, literate speech, ability to keep and switch the attention of the audience Ability to communicate with the course team Possibility to connect an assistant	Not of importance Possibility to connect an assistant
The design of the course	Nonstandard	Standard with elements

4 Common signs of all MOOCs

The list of students who have higher education depends on a number of points: from the direction to which the course belongs (art, humanities, business and management, computer technologies, biological sciences, psychology, physics, mathematics and logic, engineering sciences, etc.), from the purpose of the course (enlightenment / education, career guidance, specific skills, training, retraining, etc.), from the necessary preliminary training (if any special knowledge that you need to have for the course / prerequisite training course). Also, the same course on different platforms can have different target audiences. In courses aimed at career guidance, training, retraining, practical tasks should be more than, in popular science courses. Practical tasks are a more effective way to check the achievement of the results announced in the course (working out practical skills, checking understanding of the organization of processes, getting feedback on the chosen option, etc.). In popular science courses it is possible to organize a test at the expense of a system of tests (verification of theory assimilation), although practical assignments in such courses are perceived positively by listeners.
Common problems of MOOC using:

- the difference in the level of students' training, inadequacy of peer-to-peer evaluation;
- the problem of identification of the student;
- the inaccessibility of teacher in large numbers in groups;
- lack of binding knowledge to a specific country, its needs and features;
- ensuring a minimum level of language training;
- the need for prior focus group courses;
- unlimited access to online course content after the course ends;
- the need of adapting teaching methods to the digital culture of generations (generations Y-Z);
- the need of including courses directly in the educational programs of universities (as well as in the evaluation system as a whole;

- the difficulty of including the business community and professional communities as experts to assess the quality of online courses and post-course learning outcomes;
- addressing financial issues related to the use of courses.

Conclusion

The course analysis was conducted. The peculiarities and problems of development and implementation of online courses were revealed.

References

1. D. Badarch, N. G. Tokareva, M. S. Tsvetkova, Vysshee obrazovanie v Rossii, **10** (2014)

2. I. Ainoutdinova, A. Blagoveshchenskaya, *Proceedings of ICERI2017*, (2017)

3. Information about MASSIVE OPEN ONLINE COURSES from web-site URL:*http://iite.unesco.org/oer_and_digital_pedagogy/oer/online_courses/*

4. Watted Abeer, Barak Miri, Procedia - Social and Behavioral Sciences, **152** (2014)

5. I. Nikanorov, A. Shvindt, MATEC Web of Conferences **141**, 01054 (2017)

6. K.V. Mertins, V.S. Ivanova, A. V. Shadrina, Humanities Education Issues, CITEEMPTI, (2017)

Application of satellites system based on different heights for ionospheric disturbances monitoring

Ivan Kaloshin*, *Vladimir* Skripachev, *Irina* Surovceva, *Vladimir* Kuznetsov, and *Alexander* Kharlamov

Moscow Technological University (MIREA), 119454 Moscow, Russia

Abstract. Results of data processing on-board Langmuir probes of CHAMP and DEMETER satellites are presented. The characteristics of satellites are given. Analysis of ionospheric parameters before strong earthquake is performed. At the altitudes of the satellites anomalous changes in the ionosphere parameters were detected a few days before the seismic event. Frequency distribution of anomalous ionospheric disturbances is obtained.

1 Introduction

Earthquakes are one of the most significant disasters for the people and the infrastructure of our country. In Russia more than 25% territory and near 20 million people could be exposed to the earthquakes with magnitude $M \geq 7.0$.

A lot of empirical material and researches of earthquakes precursors before earthquakes were collected for the last decades. It is known that earthquakes precursors appear in different geospheres, namely: lithosphere, atmosphere and ionosphere.

According to [1-3], ionospheric precursors observed up to several hours or days before an earthquake. Taking into account the ionosphere's variability, monitoring ionospheric precursors by satellites at different altitudes using identical onboard equipment has certain scientific interest.

2 Data sources

German satellite CHAMP has been working from July 15, 2000 to September 19, 2010, being in orbit with the altitude of ~380 km and the inclination of 87.3°. French satellite DEMETER was launched to circular orbit with height ~700 km and has been operational from June 29, 2004 to December 9, 2010. Figure 1 shows satellites relative position and idealized electron density profiles for day (solid line) and night (dotted line) ionosphere.

Satellites data were obtained from the web sites (http://gfz-postdam.de and http://cdpp-archive.cnes.fr, respectively). Both satellites had Langmuir probes. They are one of the

* Corresponding author: kaloshin@mirea.ru

effective instruments [4] to measure: Ne - electron density, Ni - ion density, Te - electron temperature (DEMETER satellite), and Ne - electron density (CHAMP satellite). Kp indices for analysis were obtained from the information on the web site http://omniweb.gsfc.nasa.gov.

For current study we have selected 12 earthquakes. Criteria of selection were: depth of epicenter of not more than 50 km and magnitude M ≥ 6.0, occurred in the period of satellites' operational activity (http://earthquake.usgs.gov).

Table 1 shows the List of the earthquakes selected in this research.

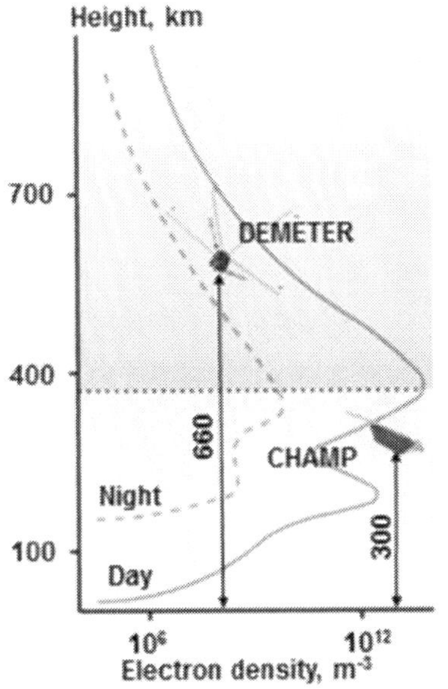

Fig. 1. Satellites relative position and idealized electron density profiles.

Table 1. List of the seismic events selected in this study.

#Case	Case study	Date	Magnitude	Geographic latitude	Geographic longitude	Depth h, km
1	Borujerd, Iran	31.03.2006	6.1	33.50N	48.78E	7
2	Kamchatka Peninsula, Russia	20.04.2006	7.6	60.95N	167.09E	22
3	Kuril Islands, Russia	15.11.2006	8.3	46.59N	153.27E	10
4	Taiwan	26.12.2006	7.1	21.80N	120.55E	10
5	Central Peru	15.08.2007	8.0	13.39S	76.60W	39
6	Antofagasta, Chile	14.11.2007	7.7	22.25S	69.89W	40
7	Xinjiang, China	20.03.2008	7.2	35.49N	81.47E	10
8	Sichuan, China	12.05.2008	7.9	31.00N	103.32E	19
9	Honshu island, Japan	19.07.2008	7.0	37.55N	142.21E	22
10	Minahasa Peninsula, Indonesia	16.11.2008	7.4	1.27N	122.09E	30
11	L'Aquila, Italy	06.04.2009	6.3	42.33N	13.33E	8
12	Samoa Islands	29.09.2009	8.1	15.49S	172.10W	18

Time period for data analysis was [-60;+10] days around earthquake. Thus, 71 days were analyzed for each earthquake.

3 Processing technique

After collecting the data available for our study, we have developed some calculation procedures to process them.

First step of processing satellite data was performed taking into account the zone of earthquake preparation, its size. Radius of earthquake preparation zone estimated using the formula $R=10^{0.43M}$, where M is the earthquake magnitude. In defined radius data were classified by local time as day (06-18LT) and night (18-06LT) time. Afterwards data have been averaged.

Next, we used well-known technique that is based on the median value and the inter-quartile range. This technique detailed in the paper [5]. For obtained time series of average values we have calculated median value, inter-quartile range. These values were used to consider low and high bounds for each parameter (for both CHAMP and DEMETER data) by following formula:

$$Bound = M \pm k \cdot IQR,$$

where M - median, k – coefficient which should be proportional to the earthquake magnitude. In this paper we used k=1.5 for all earthquake cases. All values outside bounds were considered as anomaly signals.

Figures 2 and 3 shows the results for Central Peru earthquake happened on 15.08.2007 with the magnitude M=8.0. Day 0 corresponds to the earthquake day and is marked by dark arrow. Geomagnetic conditions were quiet in general.

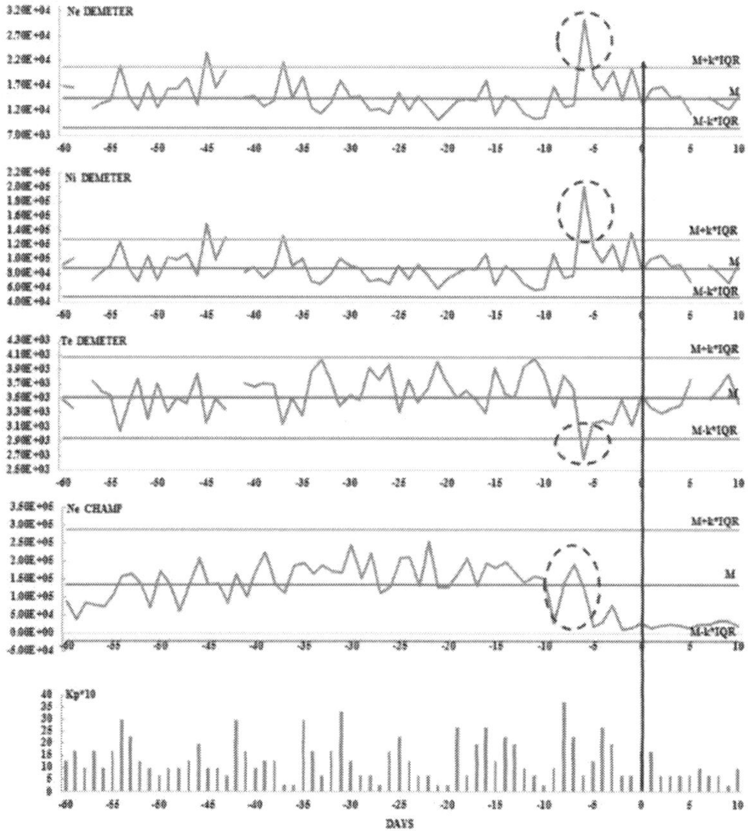

Fig. 2. DEMETER and CHAMP day time parameters and Kp indices.

Note that Ne, Ni, and Te values increased up to -7, -6 days in the day time according to DEMETER data. For CHAMP data, Ne disturbances were not detected, but there is a significant increase by -8, -7 days. During the night time data processing for the DEMETER satellite, values of Ne and Ni increased by -10 and -9 days, as well as the growth of Te by -10 and -5 days was registered. According to the CHAMP data, there was the growth of Ne values, which began at -15 days with the formation of anomalies signals up to -8, -3 and -4 days.

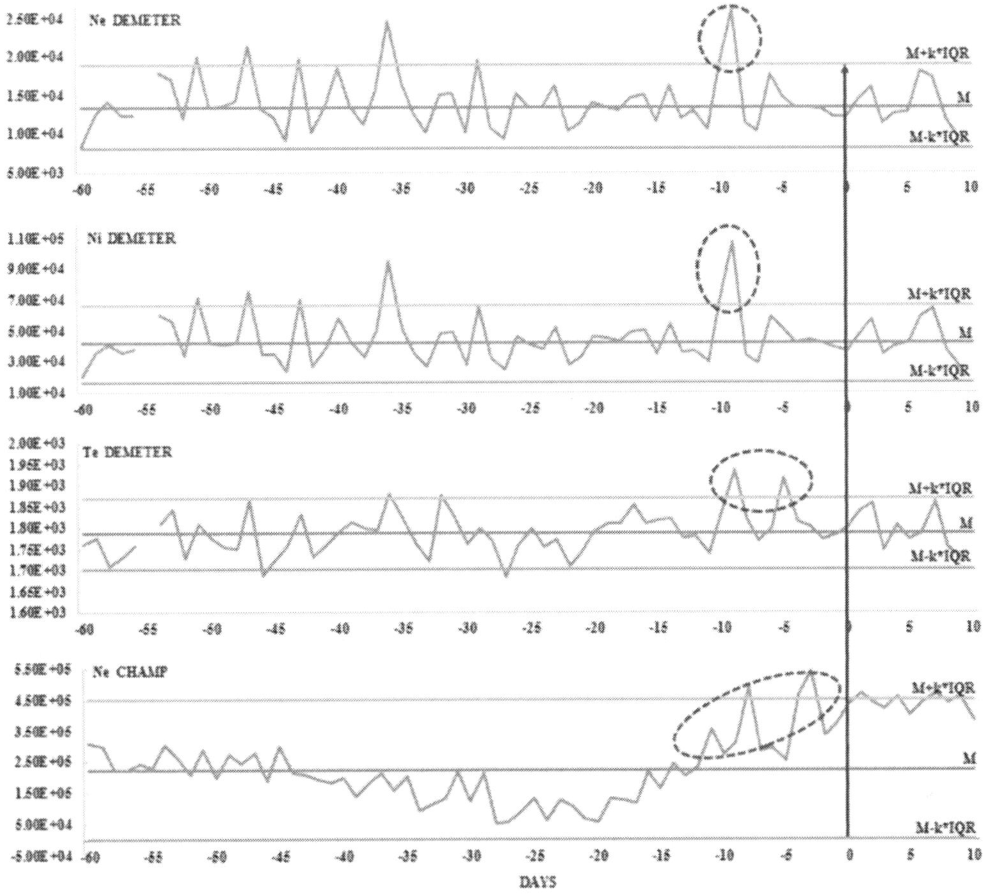

Fig. 3. DEMETER and CHAMP night time parameters.

4 Results

Described procedures have been implemented for all selected earthquakes. Results for all parameters were summarized in the Figure 4, which shows the frequency distribution of anomalous ionospheric disturbances.

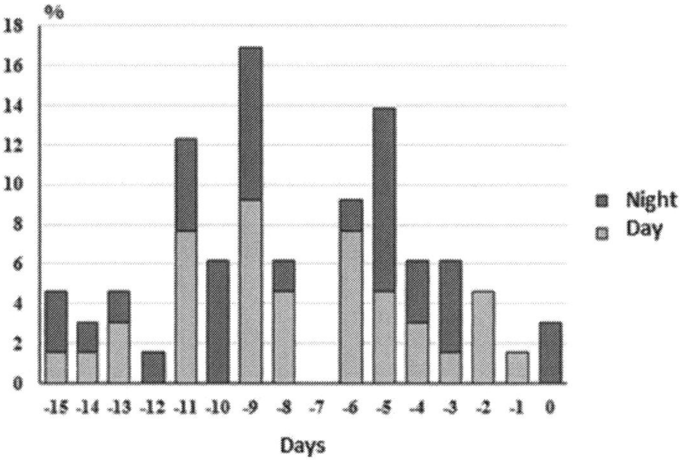

Fig. 4. Frequency distribution of appearance of anomalous ionospheric disturbances.

It should be noted that the greatest occurrence of significant variations was observed up to -9, -5 and -11 for daytime. It is noticeable also that the greatest amount of ionospheric disturbances for the studied earthquakes was observed in the period from -11 to -5 days. At first, the ionospheric disturbances, recorded during daytime, dominate, and then the predominance at night prevails. Attention is drawn to the fact that the disturbances at -7 days are absent.

5 Conclusion

Thus, we have performed the complex analysis of Langmuir probes obtained from CHAMP and DEMETER satellites. Our results are coherent with other studies, such as, for instance [4]. Simultaneous ionosphere monitoring at different altitudes could be useful by using serial small satellites.

The authors thank Centre des Données de la Physique des Plasmas (CDPP) and German Research Centre for Geosciences (GFZ) for the use of the DEMETER and CHAMP data respectively.

References

1. S. A. Pulinets, D. Ouzounov, D. Davydenko, A. Petrukhin, E3S Web of Conf. **11** (2016)

2. A. Rozhnoi, M. Solovieva, M. Parrot, M. Hayakawa, P.F. Biagi, K. Schwingenschuh, V.Fedun, Physics and Chemistry of the Earth, Parts A/B/C **85-86** (2015)

3. M. Parrot, V. Tramutoli, J.Y. Liu, S.A. Pulinets, D. Ouzounov, N. Genzano, M. Lisi, K.Hattori, A. Namgaladze, Nat. Hazards Earth Syst. Sci. Discuss. **172** (2016)

4. I. Kaloshin, V. Kuznetsov, V. Skripachev, I. Surovceva, MATEC Web of Conf. **102** (2017)

5. M. Akhoondzadeh, M. Parrot M, M.R. Saradjian, Nat. Hazards and Earth System Sci. **10** (2010)

Betatron radiography and tomography of steel castings with large thickness

Daniyar Kayralapov, *Yang* Zhong, *Andrei* Batranin*, and *Sergei* Chakhlov

Russian–Chinese Laboratory of Radiation Testing and Inspection, National Research Tomsk Polytechnic University, 634050 Tomsk, Russia

Abstract. Steel castings with large thickness are widely used in different areas of industry and the control of steel castings with large thickness is becoming more and more important in order to detect defects and ensure the reliability. In this paper, we carry out betatron radiography and tomography to control the steel castings with large thickness and check the potential ability of Inspection and Examination System (IES) for a high-energy betatron tomography. The results of betatron radiography and tomography of steel castings with large thickness are presented and compared with precedent work, which shows that the IES is considered a promising high-energy tomography system.

1 Introduction

Steel castings of a large thickness are widely used in machinery, shipbuilding, aerospace, oil and gas industry, etc. The conditions for applying steel castings with large thickness are critical: high temperature, extreme pressure, corrosive media, etc. These requirements put restrictions on production.

The examination of steel castings with large thickness is a very important work in order to detect defects and ensure the reliability. In order to detect possible shrinkage porosity, shrinkage shells, inclusions and other defects, it is necessary to set up examination by a high-energy x-ray tomography, which returns complete information of the inspected object with a large thickness. One of the key tasks of the steel castings inspection is the identification of the defect generation mechanisms. Such identification helps with producing optimization, improvement in technology of steel casting production, system reliability and hardware lifetime. Meanwhile non-destructive high-energy x-ray testing is necessary for assessment of the change in density and recognition of detrimental defects in newly produced steel castings with large thickness.

Now many companies produce and develop high-energy tomography to test the thick-walled, large-sized steel castings and products of complex construction. Foreign companies such as Fraunhofer IIS [1], Empa [2], Jesse Garant Metrology Center [3], GRANPECT [4], CZST [5], and domestic companies like PromIntro [6] and JSC "FNPTs Altai" [7]. In our case, the betatron radiography and tomography of steel castings with large thickness were carried out with the help of the Inspection and Examination System (IES) [8] at the Tomsk

* Corresponding author: batranin@tpu.ru

Polytechnic University. The IES was used to inspect a steel castings with large thickness as well a high-energy betatron tomography. By this work, we studied the potential ability of IES working as a reliable tomography system.

2 Materials and Methods

During inspection of the steel castings with large thickness, we conducted experiments of radiographic and tomographic scans with the help of IES as high-energy tomography in Russian–Chinese Laboratory of Radiation Testing and Inspection of the Tomsk Polytechnic University. The general view of the IES and the inspected object are presented in Fig. 1 below.

(a) (b)

Fig. 1. General view of the IES system (**a**) and the inspected object (**b**).

The source of high-energy x-ray radiation is a betatron with 9 MeV energy. The dose rate is 18-21 R/min at 1 meter from the focal spot. The betatron is equipped with a vertical collimator. The width of the collimated beam is approximately 25 mm in the plane of the detector. The size of the focal spot in the vertical direction does not exceed 2 mm.

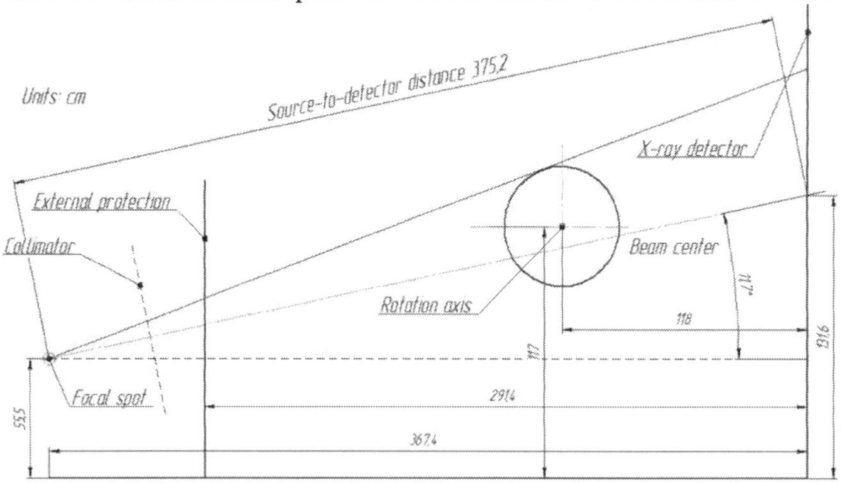

Fig. 2. Scheme of installation for betatron radiography and tomography.

The installation scheme for betatron radiography and tomography is shown in Fig. 2 on the next page. The inspected object was placed on a stand associated by a rotary shaft. The rotator is mounted on a frame with an electric drive for longitudinal displacement of the inspected object in a direction perpendicular to the radiation beam. Using the rotator allows one to shoot from different angles, as well as to collect data for the tomography examination.

For x-ray detection, we use a linear detector manufactured by TSNK-Lab (Moscow). The scintillation material of this linear detector array is $CdWO_4$ and crystal size is 5×6×30 mm^3. The detecting unit contains 608 detectors.

3 Results

Fig. 3 shows a radiogram of steel casting with large thickness and the positions of the tomography slices. Reconstruction was carried out in two programs: INKCT (developed by TPU) and Nrecon (developed by Bruker microCT, Belgium) [9]. In tomographic experiments, the number of projections for each section was 1800 with an angular pitch of 0.2 degrees. The pixel size of the tomograms is 1.72 × 1.72 mm^2. The size of the tomogram is 280 × 280 pixels. Below are the results of tomographic cross sections obtained in the experiments (Fig. 4).

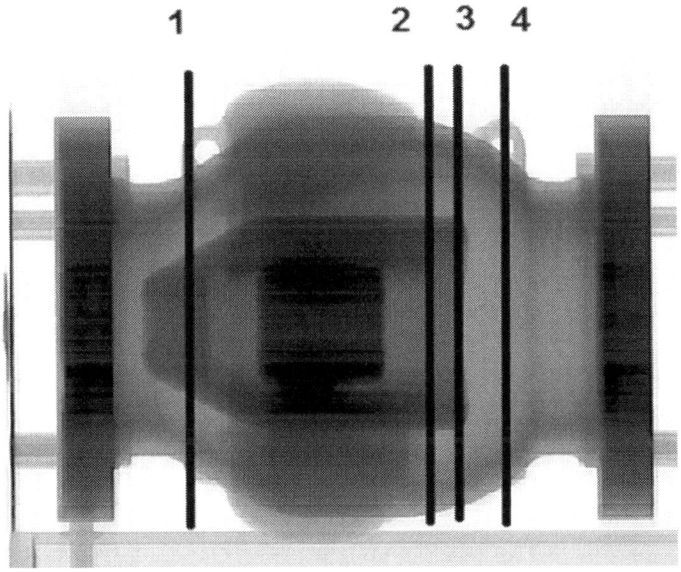

Fig. 3. Radiogram with the extraction of details at a large thickness and the position of the tomography slices.

The presented tomograms in Fig. 4 show that the inspected object is sufficiently transparent for imaging. However, as in the case of radiography, the resolution of the tomograms is insufficient to analyze the presence of defects. In addition, the experiments are burdened by the non-optimal geometry of the tomographic setup. In this case, there is a reason to state that correcting the shortcomings of the installation will provide spatial resolution at the level of 1-2 pixel sizes of the tomograms.

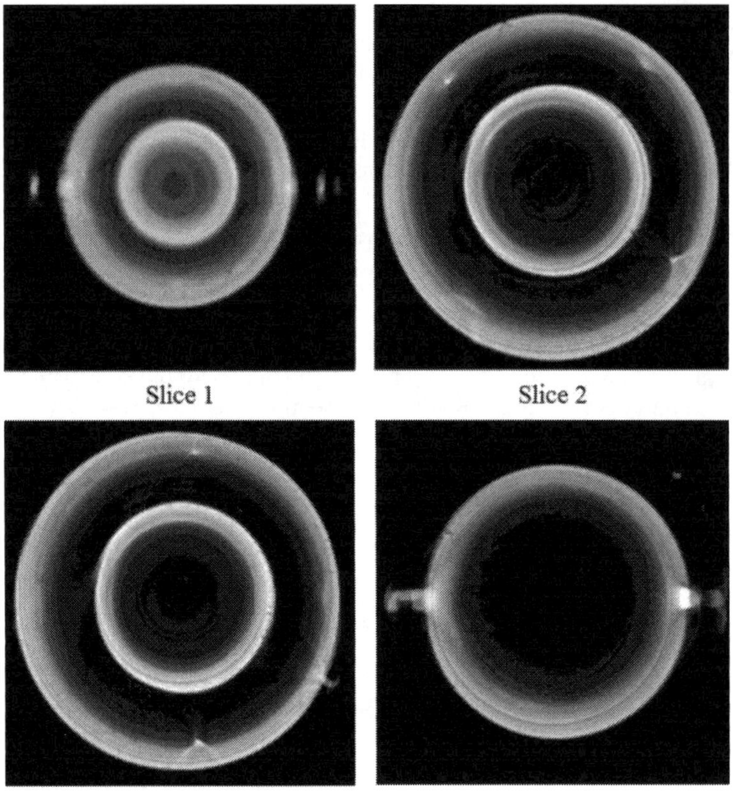

Fig. 4. Tomograms of the selected sections.

4 Discussion

As the results in [10-12] showed, it is difficult to choose a powerful source to penetrate a steel object with very large thickness. Even a linear accelerator can hardly reach a high resolution of a steel casting with such large thickness.

This work can be considered as a continuous research study of [13] and [14]. With a significant increase of thickness of the inspected object, as stated in previous section, the resolution of the tomograms become insufficient to analyze the presence of defects as the experiments are burdened by the non-optimal geometry of the tomographic setup, which indicates the shortcomings of IES to be improved. Take into consideration of the thickness of steel casting is bigger than 100 mm, the experiment results in Fig. 3 and Fig.4 can be thought as with good quality and high resolution.

5 Conclusions

The purpose of the experiments was to investigate the possibilities of using the betatron as an X-ray source for examination of the large-sized cast products. It is shown that a 9-MeV betatron possesses sufficient quantum energy to produce a signal at the declared radiation thickness. It is also shown that radiography and tomography of steel castings with large thickness are possible in principle. However, when designing an installation, it is necessary to strive to increase the resolution of the obtained images by including as many possible elements of the detector as possible. At the same time, the volume of the scintillator should not decrease significantly. In spite of the shortcomings of IES, this work shows the potential ability of IES working as a promising high-energy tomography system.

References

1. High energy CT or XXL-CT. URL : https://www.iis.fraunhofer.de/en/ff/zfp/tech/hochenergie-computertomographie.html (2018)
2. X-ray Computed-Tomography. URL: https://www.empa.ch/web/s499/x-ray-ct (2018)
3. 3 MeV Computed Tomography System. URL: http://www.prweb.com/releases/2017/09/prweb14750508.htm (2018)
4. IPT Series of Industrial CT NDT System. URL: http://www.granpect.com.cn/english/products/p1/27.html (2018)
5. High energy CT. URL: http://www.meansee.com/index.php?m=product&a=index&cid=11 (2018)
6. E.I. Vainberg, I.A. Vainberg, In the world of non-destructive control, **3**, 18-23 (2013)
7. Tomographic methods and tools. URL: http://frpc.secna.ru/def/def6.php (2018)
8. Inspection and Examination System. URL: http://portal.tpu.ru/departments/laboratory/rknl/products/IDK (2018)
9. Bruker microCT | Software downloads. URL: http://www.bruker-microct.com/products/downloads.htm (2018)
10. Y.S. Xiao, H.F. Hu, Z.Q. Chen, L. Zhang, T. Yong, CT Theory and Applications **18**, 72-78 (2009)
11. M. Salamon, M. Boehnel, N. Reims, G. Ermann, V. Voland, M. Schmitt, N. Uhlmann, R. Hanke, *5th International Symposium on NDT in Aerospace*, Singapore (2013)
12. Y.S. Xiao, Z.Q. Chen, Y.T. Li, L. Ye, *19th World Conference on Non-Destructive Testing* (2016)
13. A.M. Shteyn, A.I. Cheprasov, S.V. Chakhlov, M.M. Shteyn, V.A. Klimenov, S.V. Knyazev, *18th World Conference on Nondestructive Testing* (2012)
14. Y. Zhong, S.V. Chakhlov, T. Mamyrbayev, MATEC Web of Conferences **102**, 01038 (2017)

MATEC Web of Conferences 158, 01017 (2018) https://doi.org/10.1051/matecconf/201815801017
VI International Forum for Young Scientists "Space Engineering 2018"

A search and improvement of the geometric parameter betatron injector

Andrey Kolomeytsev[1,*], and *Mikhail* Shtein[2]

[1]National Research Tomsk Polytechnic University, Department of Electronic Engineering, 634050 Tomsk, Russia
[2]National Research Tomsk Polytechnic University, Research and Production Laboratory "Radiation Inspection and Industrial Safety Systems", 634050 Tomsk, Russia

Abstract. The article describes the search for optimal geometric parameters of the betatron injector. First of all, we determine parameters that influence the injection efficiency. Next, we introduce the coefficient that directly connects the geometric parameters of the injector with capture efficiency. The computer model of the betatron injector is recreated. In the injector region, there are forces that are equal in magnitude and direction to forces in a real betatron. We simulate different parameter combinations. After processing the results, tolerances are obtained for the selected parameters. The results of this work can be used as recommendations for the development of design documentation for new three-electrode injectors for betatrons. The research method can be used to study the efficiency of other injector designs, as well as to search for new design alternate.

1 Introduction

The injector is the key betatron component. Injector parameters determine the maximum number of electrons that can be captured by the magnetic field of a betatron. However, the optimal geometric parameters of an injector have not been defined up to the present day. The manufacturing tolerance of these parameters is still undefined either. From the large number of electrons emitted by a cathode injector, only a small percent is captured by the accelerating field. There are many reasons behind this phenomenon that are confirmed experimentally. These include the following:

- Insufficient injection energy;
- Low electron space-charge density;
- Low synchronization accuracy between the injection and accelerating flux;
- Insufficient stability of the voltage pulse generation;
- Non-optimal geometry and relative position of electrodes [1-5].

This research work considers how geometrical parameters of the injector affect the coefficient of its current transmission.

The active part of the capture process takes approximately one microsecond. During this period of time, the initial electron velocity (injection voltage) is adjusted to the magnetic

* Corresponding author: aak65@tpu.ru

field magnitude. Meanwhile, (4…5) $\cdot 10^{12}$ electrons are injected into the interpole space. A betatron only captures (1…1.5) $\cdot 10^{10}$ particles, in other words, one electron out of 300...500. The ratio of the captured electrons to the injected ones is capture efficiency [6-8].

2 Materials and methods

Betatron electromagnet has an annular gap and is supplied with an alternating current. The accelerating chamber is set in the annular gap. The chamber has a high vacuum (10-6 mm. Hg. Art.) inside. The injector is mounted in one of the accelerating chamber nipples.

Figure 1 shows the image of the injector and its geometrical parameters.

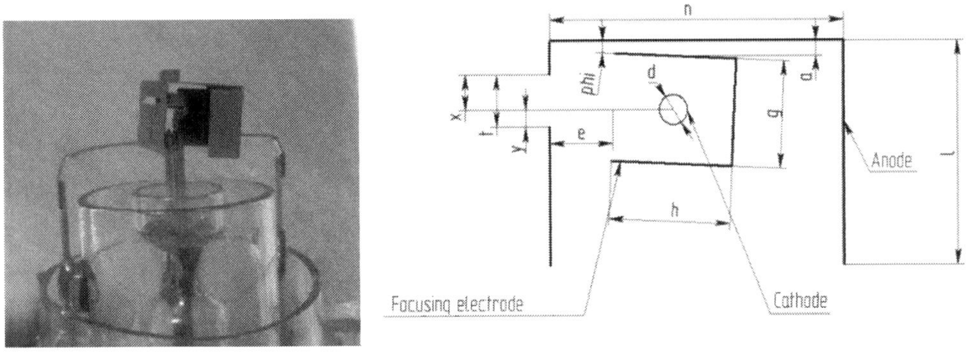

Fig. 1. The injector and a schematic drawing of labeled parameters.

The parameter a determines the electrical injector stability and shows the distance between the center of the top plate of the focusing electrode and the anode.

The parameter e shows the distance between the focusing electrode and the anode gap.

The parameter f (not shown in the figure) is called misalignment and is defined as follows. Through the middle of the focusing electrode a projection extends perpendicular to the gap line. Next, we measure the distance from the line to gap edges x and y, and calculate the misalignment f according to formula (1). The displacement to the betatron center is positive.

$$f = \frac{x-y}{2} \quad (1)$$

The angle of the focusing electrode against the anode φ is the angle between the top of the anode and the focusing electrode plate. A counterclockwise rotation is positive.

The parameter t is the anode gap width. Electrons leave the injector through it.

The parameters h and g determine the width and height of the focusing electrode, respectively. It is also necessary to take into account the horizontal sH and vertical sV displacement of cathode relative to the focusing electrode center. Of particular importance is the parameter d that describes the cathode diameter.

The dependence of the coefficient of the current transmission on the geometric parameters of the injector is determined by multi-objective optimization. To do this, we set the values of all the parameters to certain (factory) values, then we change only one of them and observe how it affects the result.

Currently, there are several theories that describe the electron capture mechanism in acceleration. However, they only give a qualitative description. So, now it is not possible to create an analytic model to directly connect the geometric parameters of the injector with capture efficiency. However, there is a parameter that can be included in the analytic model to make capture efficiency easy to measure. This is called the coefficient of the current transmission of the injector.

$$K_{\mathrm{T}} = \frac{I_L}{I_A + I_L} = \frac{I_L}{I_K} \tag{2}$$

As be seen from the formula (2), the maximum injector efficiency is achieved when the coefficient of the current transmission equals. Based on this statement, the main purpose of this study is to define tolerances of the injector parameters, at which the value tends to 1 [9].

In this case, the current transmission coefficient is the ratio of the total charge of the electrons emerging from the injector to the total charge of the electrons that overcome the accelerating gap of the injector, some of which face the anode.

$$K_{\mathrm{T}} = \frac{N_L}{N_A + N_L} = \frac{N_L}{N_K} \tag{3}$$

where - N_L - is the total charge of the electrons that overcome the accelerating gap of the injector, N_K – is the total charge of the electrons emerging from the injector.

3 Results and discussion

Inside the injector, an electron moves under the influence of the electrostatic field and the magnetic field. The magnetic field shifts an electron to the equilibrium orbit. The intensity of the electrostatic field **E** at any point is a physical quantity defined by the force acting on a charge placed at that point of the field. Intensity **E** coincides with the direction of the force acting on the charge.

Figure 2 shows the electric field potential in the central section of the injector.

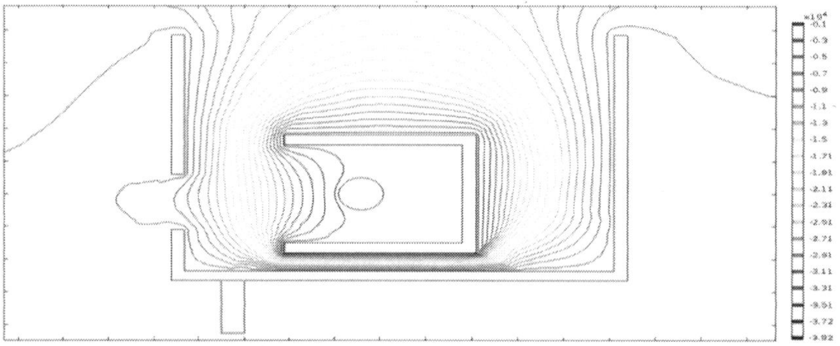

Fig. 2. Equipotential lines of the electric field.

At the injection moment, contractor biases the magnetic field in such a way as to dislocate the equilibrium orbit close to injector. It allows avoiding significant radial oscillation at the acceleration beginning. There is an empirical formula (4) for the calculation of the magnetic induction at the any radius of interpolar space. It allows calculating the required induction of the magnetic field to hold the electrons of certain energy [10].

$$(W(W + 2E_0)^{\frac{1}{2}} = 300B_0R_0 \tag{4}$$

where R_0 – is the radius of the equilibrium orbit, m; B_0 – is the magnetic induction at the equilibrium orbit, T; E_0 – is the electron rest energy, MeV.

From the knowledge of radius, electron rest energy and the fact that leaving electrons have an energy of about 40 keV, we can calculate the value of the magnetic induction at the injector.

$$B_0 = \frac{(W(W + 2E_0)^{\frac{1}{2}}}{300R_0} = 0.0064\,T$$

The next step is to simulate the motion of a charged particle in an electromagnetic field. Modeling is based on the calculation results of the magnetic and electric fields. The calculation is only possible after the imposition of boundary conditions on the task. It is important to note that the problem has a number of assumptions and simplifications. First, the electron injection is carried out at a thermionic emission cathode with heating. The average energy of the electrons leaving the metal surface with a thermionic emission $E_{TE} = 0.17$ eV. Second, the electrons can leave the cathode surface in any direction. We set the initial velocity vector directed normal to the surface of the cathode. Third, at this stage of the work, we do not take into account the Coulomb interaction of the electrons with each other because of low charge density.

Certain electron portion falls onto focusing electrode or anode (Figure 3). It means so geometrical parameters of the injector exist that each or the most electrons is able to leave the injector. The determination of such a configuration makes it possible to increase the number of successfully injected electrons and the effective cathode area.

a)

b)

c)

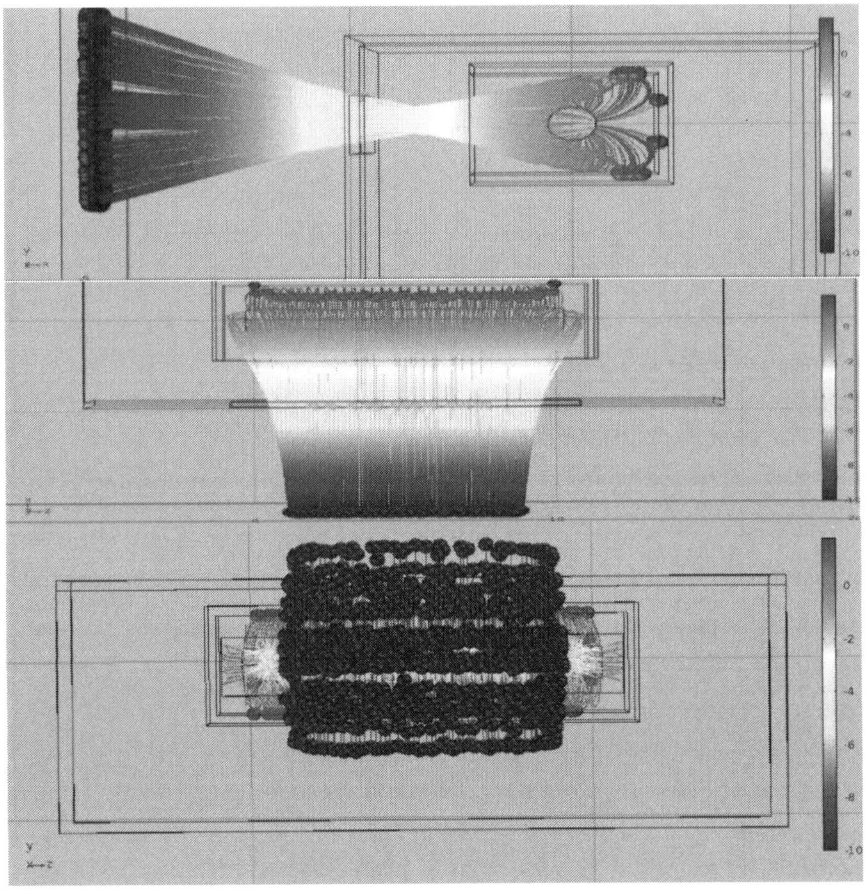

Fig. 3 (a – XY plane, b – XZ plane, c – YZ plane). Injected electron trajectories.

We have simulated acceleration with different parameter values. The results are shown in the Table 1. Table 1 describes how to set injector elements to achieve high injection efficiency.

Table 1. The summary tabulation of the injector parameters.

Parameter	Quantity	Upper limit	Lower limit
a, mm	0.7	0.9	0.55
e, mm	2.2	3.0	-
f, mm	0.15	0.4	0
t, mm	1.75	1.9	1.5
φ, deg	0	3.5	2.5

Parameter	Quantity	Upper limit	Lower limit
g, mm	3.6	3.8	3.4
h, mm	4.2	-	3.9
sH, mm	-0.4	-0.2	-0.5
sV, mm	0	0.4	-0.3

4 Conclusion

In this study, we have identified the most important parameters, which have an effect on electron injection. A computer model is used to simulate some of the processes inside the injector. The paper has shown how the geometrical parameters of the injector affect the coefficient of the current transmission. We have also calculated the optimal values of these parameters and assigned the appropriate tolerances to them. This set of parameters allows us to obtain an injector whose current transmission will always be $K_T \geq 0.9$. The results of this work can be used as recommendations for the development of design documentation for new three-electrode injectors for betatrons. The research method can be used to study the efficiency of other injector designs, as well as to search for new design alternate.

References

1. J. T. Seeman, *Proc. of B FACTORIES* **233** (1998)
2. H. Wiedemann, *Particle Accelerator Physics I – Basic Principles and Linear Beam Dynamics* (Springer) ISBN 3-540- 64672-X
3. V. Moskalev and V Chaklov *Betatrons* (Tomsk: TPU Publish House, 2009) in Russian
4. M. Aiba et al., Phys. Rev. ST Accel. Beams **18**, 020701 (2015)
5. D Greene, The British journal of radiology **34(398)** (1961)
6. V Moskalev and G Sergeev *The Induction Electron Accelerator – Betatron* (Tomsk: TPU Publish House, 2012) in Russian
7. Rychkov M. M. et al. J. of Phys.: Conf. Ser. **881(1)**, (2017)
8. I. Zatonov, M. Shtein, MATEC Web of Conf. **48** (2016)
9. Shestak A. P. et al., IOP Conf. Series.: Mat. Sci.&Eng., **81(1)**, (2015)
10. A. Sedoi and M. Shtein, Instruments and experimental techniques **19(5)**, (1976)

MATEC Web of Conferences **158**, 01018 (2018) https://doi.org/10.1051/matecconf/201815801018
VI International Forum for Young Scientists "Space Engineering 2018"

Results of experimental studies and numerical modeling of multistage waveguide-slotted membranes filters with complex slots geometry

Natalia Kopylova[1,*], *Alexei* Kopylov[1], and *Yuri* Salomatov[1]

[1]Siberian Federal University, 660079 Krasnoyarsk, Russia

Abstract. The article contains the results of an experimental study and numerical simulation of the amplitude-frequency characteristics of one, three and four-stage filters on waveguide-slotted membranes with complex resonance slot geometry membranes located in a rectangular 35x15 mm waveguide. The design of filters and the topology of a complex slot in the membrane are shown. The results can be useful in the design of waveguide filters in the frequency range 5.5 ... 8.5 GHz.

1 Introduction

As is known [1], carrying in regular rectangular waveguide inhomogeneity of one or another type is the one of the main ideas for realization microwave filters based on rectangular waveguides. In largest number cases, the reactive pins located along the axis of symmetry of the wide wall of the waveguide in the direction of propagation of the microwave electromagnetic wave are used as inhomogeneities. The most significant drawback of such filters, in our opinion, is their relatively large dimensions, since the distances between reactive pins should be commensurate with a quarter of the microwave wavelength at operating frequencies.

As was shown earlier [2], one of the possible ways to reduce the dimensions of waveguide filters is the formation of band-pass and band-stopping amplitude-frequency characteristics (AFC) of these filters by using thin metal membranes in their structure having resonant slots of various shapes (topology, geometry). In particular, the geometry of such slots varies from simple linear shapes [2] to angular ones located at different points in the cross-section of a rectangular waveguide [3].

Varying the geometry and location of the resonant slots allows one to achieve an increase in selectivity and obtain an AFC of a certain form. In addition, in order to increase the selectivity in such filters, a well-known cascading technique is used [2, 3]. In this case slotted membranes are cascaded with waveguide inserts of different working cross-sections and thicknesses.

In this paper we present the results of an experimental study and numerical simulation of three filters on waveguide-slotted membranes (WSM) having complex shape located in a rectangular waveguide with a working cross-section of 35x15 mm.

* Corresponding author: kopaph@yandex.ru

2 Filters design

The structures of the first, second and third filters are shown in Figure 1a), b), c), respectively.

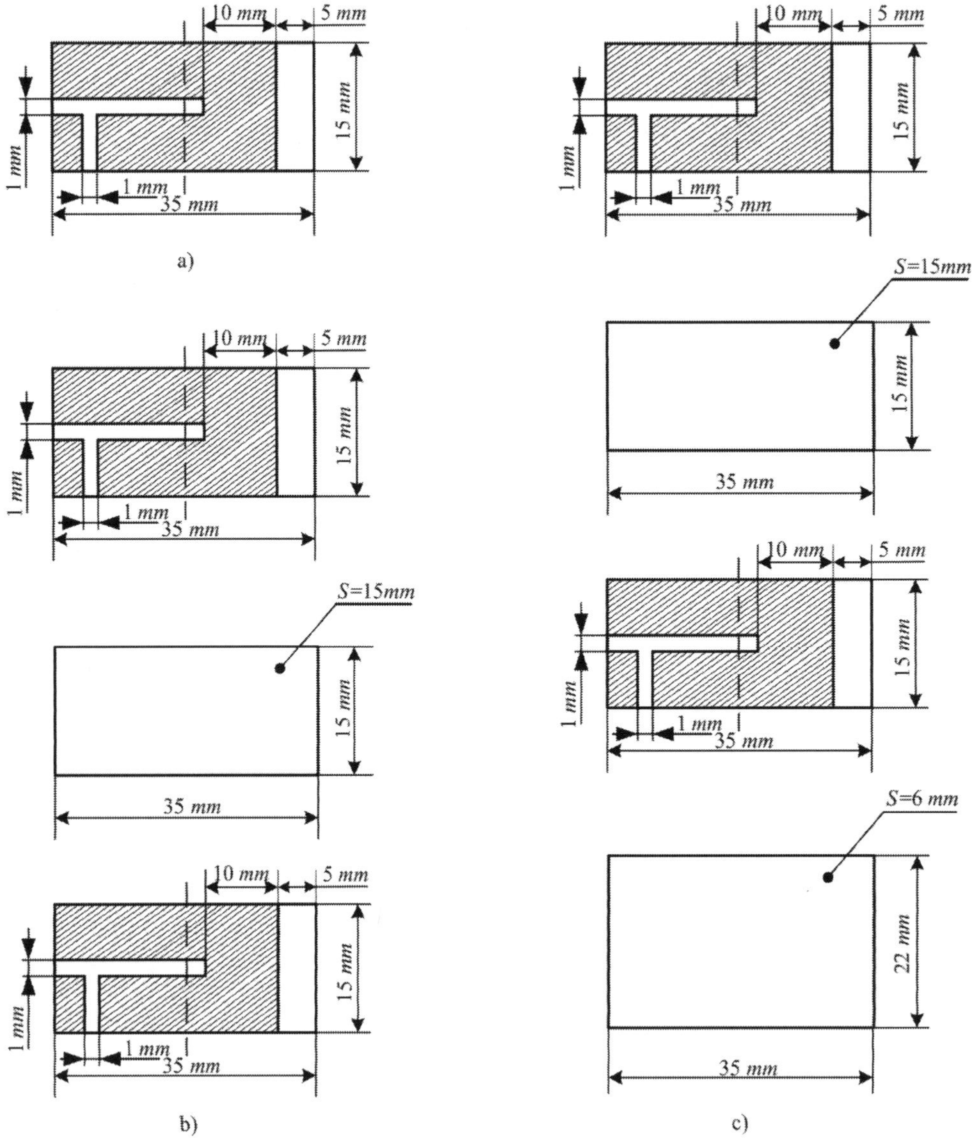

Fig. 1. The structure of the filters: a) the first filter, b) the second filter, c) the third filter

The first filter contains a single membrane with a complex geometry slot. This slot membrane is clamped between input and output waveguide-coaxial transitions (WCT) and is located perpendicular to the propagation direction of the microwave wave. The thickness of the membrane is 0.25 mm.

The second filter is a three-stage system and consists of a first single slotted cascade, a waveguide spacer S = 15 mm thick with a cross-section of 35x15 mm and a second single slotted membrane. The geometries of the slots at the slot membranes of this filter are similar to the geometry of the slots at the slot membrane of the first filter. The second slot membrane positioned symmetrically to the first membrane.

The third filter is a four-stage and consists of a first single slotted cascade, first waveguide spacer S = 15 mm thick with a working cross-section of 35x15 mm, a second single slotted membrane with a slot geometry similar to the geometry of the first membrane and arranged symmetrically to the location of the first and second waveguide spacers S = 6 mm thick with a working section of 35x22 mm.

The second and third filters, like the first filter, are clamped between input and output WCT for experimental studies and are perpendicular to the direction of propagation of electromagnetic waves in the waveguide.

The topology of the slotted membranes was chosen experimentally to obtain AFC filters closest to rectangular band-pass characteristics in the range 6 ... 7 GHz. Designs of the second and third filters with waveguide inserts are chosen experimentally.

3 Results and Discussions

Measurements of the frequency response of the filters were carried out on a scalar attenuation meter (voltage transmission coefficients, or voltage gain | K_U |) and VSWR in the manual mode of frequency tuning by points. The Figure 2, Figure 3, Figure 4 shows the results of measurements of the values of the transmission coefficients for voltage | K_U | in dB by voltage (vertical axis of the figures) against the corresponding frequencies (horizontal axis of the drawings) for the first, second and third filters, respectively. The experimental values are denoted by circular points that are connected by a solid curve.

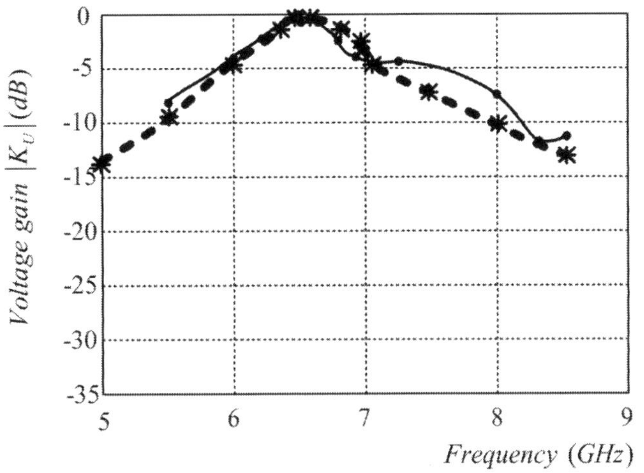

Fig. 2. Amplitude-frequency characteristics of first filter

For the same three filters was made simulation AFC by known numerical methods of electrodynamics. Results of this simulation in the form of asterisks are plotted in the same Figure2, Figure 3, Figure 4 corresponding to the first, second and third filters; the asterisks in each figure are connected by a dashed curve.

Analysis of the obtained experimental frequency dependences of | K_U | for all three filters shows that response these filters are passband by the type of frequency. At the same time, the frequency-selective properties of the first filter containing only one slot membrane providing attenuation beyond the pass band to 12 dB at a frequency of about 8.35 GHz and 8 dB at a frequency of 5.5 GHz. It seems to us satisfactory given the insignificant dimensions of the filter itself on such a single membrane - the dimensions of the filter are almost equal to the dimensions of the two WCR - input and output. The minimum losses of the first filter were 0.45 ... 0.8 dB in the pass band 6.4 ... 6.7 GHz.

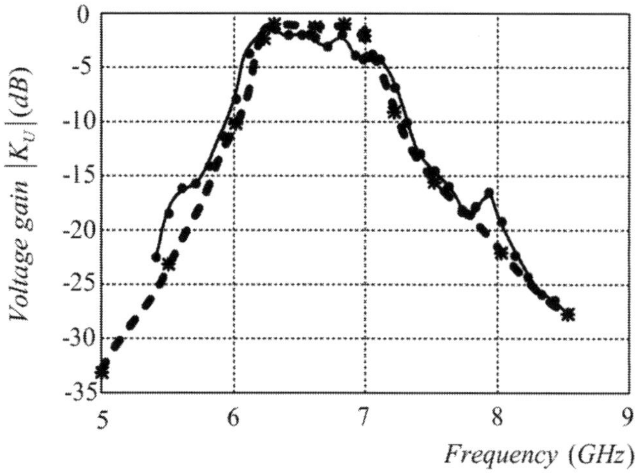

Fig. 3. Amplitude-frequency characteristics of second filter

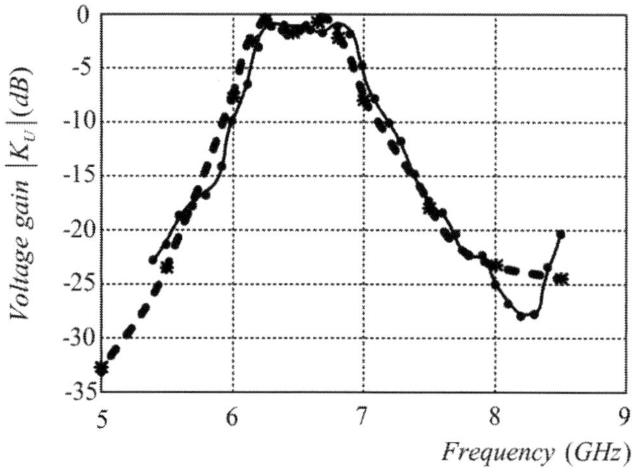

Fig. 4. Amplitude-frequency characteristics of third filter

With an increase of number filter stages, the attenuation beyond the pass-band also increases, the bandwidth of the transmitted frequencies decreases, and the slopes of the amplitude-frequency characteristics increase. So, for the second, three-stage filter, the attenuation beyond the pass-band reaches a value of 23 dB at a frequency of 5.5 GHz, 27 dB at a frequency of 8.5 GHz. At the same time, the minimum losses increase to 0.8 ... 1.8 dB in the 6.2 ... 6.6 GHz band. We can see, the AFC of the second filter shifts in frequency to the region of lower frequencies with respect to the AFC of the first filter by about 0.2 GHz. A visual comparison of the AFC of the second filter with the frequency response of the first filter shows a sharp increase in the steepness of the slopes of the amplitude-frequency characteristic of the three- stages filter with respect to the single- stages filter.

For the third, four- stages filter, attenuation beyond the pass-band reaches 23 dB at a frequency of 5.5 GHz, 27 dB at 8.2 ... 8.3 GHz with a minimum loss of 0.65 ... 1.4 dB in the frequency band 6.25 ... 6.85 GHz. Apparently, the AFC of the third filter is practically not shifted in frequency to the region of lower frequencies with respect to the second filter AFC. Moreover, the frequency bandwidth of the third filter has become larger than the bandwidth of the second filter at the same level of loss in this band. Visually, there is a noticeable increase in the steepness of the slopes of the amplitude-frequency response of the four-link filter relative to the three- stages filter. It should be noted that in the frequency response of the third filter there is a feature regarding the frequency response of the first and second filters - there is a sharp decline of AFC curve at 8.2 ... 8.3 GHz to value of attenuation about 28 dB with followed sharp increase of AFC curve at 8.5 GHz to value of attenuation about 20 dB. We believe that this is evidence of the next pass-band appearance.

4 Conclusions

The greatest differences between the values obtained experimentally and by numerical simulation are observed in the parameter of the lowest attenuation value introduced by the filters in the pass-band. The experimental values of the losses introduced by the filters in the pass-band are 0.45 ... 1.8 dB for the filters we studied, whereas the calculated values of this parameter are 0.05 ... 0.5 dB. In the parameter of the insertion loss value introduced by the filters behind the pass-bands, in the stop-bands, there is a practical equality of the numerical values between the experimental and the calculated values. The experimental values of the insertion loss by the filters in the stop bands amount to 23 ... 27 dB for the three-stage and four-stage filters, and they are practically equal to the calculated values.

This behavior of these parameters is explained by the presence of losses, the values of which are not taken into account at the moment in our modeling. The presence of additional losses just leads to a noticeable increase in the losses in the pass bands and to a certain decrease in the attenuation values in the stop bands, which become noticeable only at very large attenuations of the order of 50 ... 70 dB or more.

The carried out researches show that qualitatively the frequency response obtained by the experimental and calculated paths coincides almost completely and has some insignificant quantitative differences. The obtained results allow speaking about the possibility of confident modeling of WSM-structures for the purpose of creating efficient microwave filtering devices.

References

1. R. V. Snyder, A. Mortazawi, I. Hunter, S. Bastioli, G. Macchiarella, K. Wu, IEEE Trans. on MTT **63**, 10, (2015)

2. C. A. Leal-Sevillano, J. R. Montejo-Garai, J., A. Ruiz-Cruz and J. M. Rebollar, IEEE Trans. on MTT **64**, 3, (2016)

3. A. Kopylov, LAP LAMBERT Academic Publishing; ISBN: 978-3-659-67685-7, (2015)

Improved system for identifying biological tissue temperature using electrical impedance tomography

Evgeniy Korolyuk*, and *Konstantin* Brazovskii

National Research Tomsk Polytechnic University, 634050 Tomsk, Russia

Abstract. This paper proposes a cheap and compact medical system that determines the temperature of an object using broadband impedance tomography. This system can be used in medicine to visualize ice structure in tissue during cryosurgical operations, as well as for fault diagnosis and location in studied industrial objects. These effects are achieved by measuring electrical impedance between electrode pairs in the measuring chamber. The assembled prototype is compact, consumes little power, and allows to non-invasively determine the impedance of a target object in real time. The research included experimental studies to determine the dependence of the impedance spectrum of saline water and muscle tissue on temperature in broad band spectrum, which allowed to obtain the dependence of total electrical impedance of target objects on temperature.

1 Introduction

One of the main problems in the medical use of freezing is the difficulty of determining the boundaries and depth of cryotherapy. Due to individual characteristics of tissue, different cooling rates, and high temperature gradients during freezing, it is not always possible to accurately determine freezing depth [1].

However, regardless of the apparent simplicity of the method, the procedure of cryodestruction does not guarantee the complete removal of abnormal tissue. Insufficient influence intensity leads to local and systemic complications, which is a direct indication for reoperation. The main cause of complications is incomplete removal of abnormal biological tissues due to insufficient intensity of cryotherapy. The complexity of correct choice of cryotherapy mode is conditioned by the fact that methods of assessing depth and time of freezing are insufficiently accurate and largely subjective. The first reason of high error probability lies in the broad range of difference in water content and electrolyte concentration in tissues of different species, which leads to significant differences in physical properties (heat capacity, thermal conductivity, and freezing temperature). The second reason is conditioned by a significant temperature gradient between the surface and internal areas of biological objects [2].

* Corresponding author: esk13@tpu.ru

It is obvious that increasing the accuracy of determining freezing depth requires more advanced methods. One of the most promising methods of determining the degree of tissue damage and cryonecrosis is measuring the active component of tissue impedance. Freezing of aqueous solutions in cells leads to a sharp increase of electrical impedance, which is a reliable indicator of the quality of cryodestruction process [3].

This paper describes the work carried out to design a low-cost medical device that uses electrical impedance tomography to determine the temperature of a target object and allows to visualize the formation of ice structures within the tissue.

2 Related work

At the moment, there exist several visualizing methods of diagnostics that allow to construct images of internal environments of a biological object. The main types of such imaging are the methods of magnetic resonance imaging (MRI) and computed tomography (CT).

2.1 Magnetic resonance and computed tomography.

The magnetic resonance tomography method is a method of obtaining images of biological objects using the phenomenon of nuclear magnetic resonance. Modern magnetic resonance scanners allow to non-invasively study internal organs, as well as various processes inside them.

The computed tomography method is a method of obtaining images of biological objects by determining the difference of attenuation of X-ray radiation depending on tissue density. Just like magnetic resonance imaging, this method allows to study internal organs and various processes inside them.

However, every method has its limitations and drawbacks. The intense vortex magnetic field during magnetic resonance scanning imposes extreme requirements to the scanner and other equipment in the scanning room, as well as to the room itself. The principle of computed X-ray tomography is based on using X-ray radiation. This results in the possibility of a patient contracting cancerous diseases.

It should be noted that given the ample capabilities of MRI and CT (both of these methods provide the most comprehensive assessment of a patient's organism condition), these procedures are still more expensive compared to conventional X-ray or ultrasonography. High costs of diagnostics, the necessity to carry it out in real time, and use of metal tips in virtually all contemporary cryotherapy tools make the use of MRI and CT impractical for high-speed cryotherapy.

In order to eliminate the limitations described above, this study employs another less costly method that does not require a special room and maintenance staff. Most of the drawbacks and limitations described can be using another method of tomographic imaging: electrical impedance tomography.

2.2 Electrical impedance tomography

Electrical impedance (or simply impedance) tomography is the name for a technique of obtaining body section images by means of electrical sounding, calculations and algorithms of reconstruction of impedance distribution (resistance of different organs to electric current).

This scanning method stood out in the late 80ies – early 90ies of the past century. The essence of the method lies in the phenomenon that when alternating electric current passes through a biological object, evoked potentials are registered on its surface. Spatial distribution of electrical conductivity inside the target object can be identified if the value of the applied current and the potential registered on the object's surface are known.

3 Implementation

3.1 Electrodes and measuring chamber

In order to carry out experiments, a measuring chamber was constructed that houses a target object. Recording electrodes are located on the internal faces of the measuring chamber as shown in Figure 1 (right). The electrodes are made of silver foil coated with AgCl (silver chloride). The choice of silver-chloride electrodes is conditioned by their low level of polarization and baseline drift, low noise level, and impedance to frequency stability.

3.2 Control board

The main element of the control board designed during this study (Figure 1, left) is an STM32f1 microcontroller built on a Cortex M3 core and working with the frequency of 72 MHz. The DMA mode was used to ensure the high speed of digital-to-analog converter (DAC) operation. A 14-bit AD9764 DAC by Analog Devices was chosen as the source of scanning current [4]. Due to high output impedance of the DAC (around 100 kOhm) and to ensure the maximum voltage linearity between the DAC output and target object, it is necessary to use a current repeater with broad operating band of output voltage and higher output impedance. An E20-10 high-speed analog-to-digital converter (ADC) module was used to measure the electric potential difference between the electrodes in the measuring chamber. This module is a four-channel, 14-bit ADC with the maximum conversion frequency of 10 MHz.

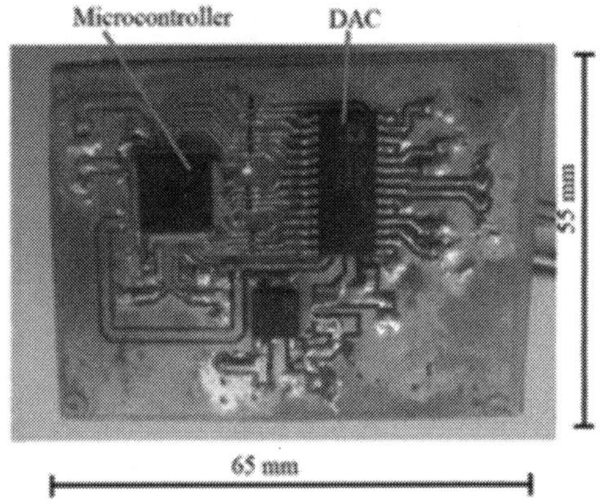

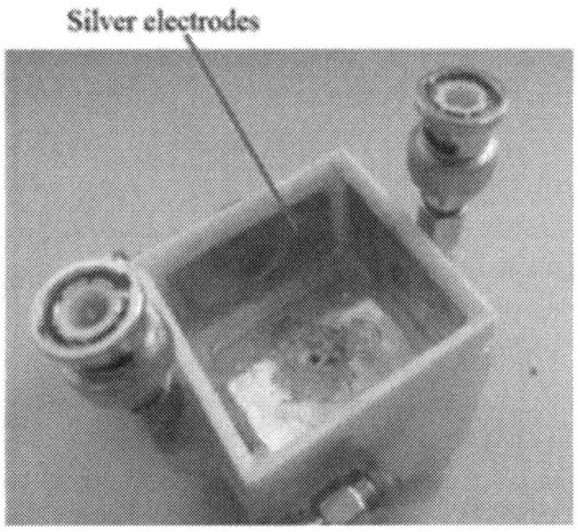

Fig. 1. Control board (left) and measuring chamber (right).

4 Results and discussion

In order to check the assembled device, it was tested using resistors with the resistance range of 50 to 500 Ohm. It is obvious that with the increase in resistance, the voltage in a resistor further drops and, consequently, power spectral density increases. The obtained results were united into a single chart as shown in Figure 3.

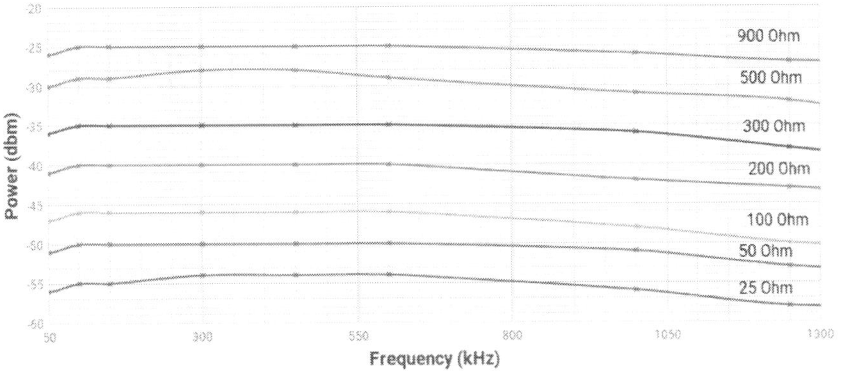

Fig. 2. Power spectral density of fixed-value resistors. The current strength is 1 mA (amplitude value). The Y axis is the power in milliwatts in the logarithmic scale of the measurement; the X axis is the frequency in kHz.

4.1 Bioimpedance measurement

Further experiments were dedicated to studying the dependence of bioimpedance on temperature in a broad band range (from 50 kHz to 2500 kHz). Target samples were represented by water with dissolved salts and muscle tissue. Temperature in the target samples varied in different range from 20 below zero to 25 above zero degrees Celsius as shown in Figures 3 and 4.

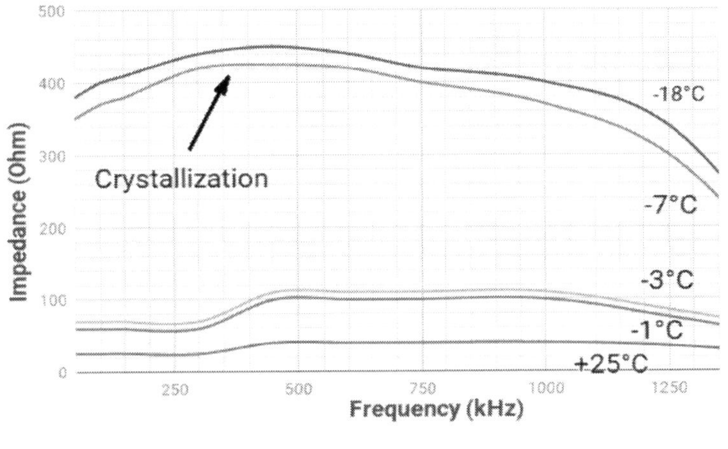

Fig. 3. Bioimpedance spectrum of water with dissolved salts with temperature change. The current strength is 1 mA (amplitude value). The Y axis is the impedance in Ohm; the X axis is the frequency in kHz.

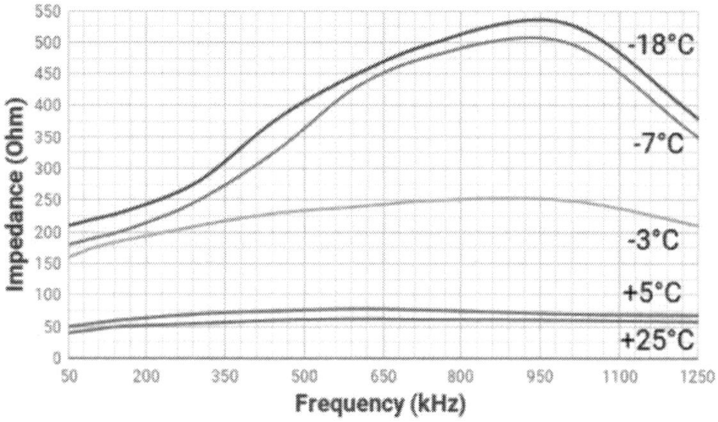

Fig. 4. Bioimpedance spectrum of muscle tissue with temperature change. Current strength: 1 mA (amplitude value). The Y axis is the impedance in Ohm; the X axis is the frequency in kHz.

4 Conclusion

This study presented a laboratory prototype of a device that uses electrical impedance tomography to identify temperature and impedance of a target object. Measurements are carried out using five electrodes located in the measuring chamber. After performing 30 measurements between two pairs of electrodes, it became possible to estimate the impedance of the target object. The experimental studies resulted in obtaining dependencies of the bioimpedance spectrum on the temperature of aqueous solutions of mineral salts and muscle tissue in the temperature range of -18 to +25 degrees Celsius. Using reference values of resistance, the accuracy of the prototype could be tested. The obtained results show that decrease of temperature leads to increase in impedance in the frequency range of 50 to 1300 kHz for aqueous solution of mineral salts and for muscle tissue. Peak values are located in the frequency of 375 kHz for aqueous solution and 950 kHz for muscle tissue. The data collected allow to reliably identify the degree of biological material freezing by measuring electrical impedance in the frequency band from 50 to 1300 kHz.

References

1. P. Laugnier, G. Berger. Assessment of echography as a monitoring technique for cryosurgery. Ultrason Imaging **15(1)**, 14 (1993)
2. J. F. Edd, L. Horwitz, and B. Rubinsky, Temperature dependence of tissue impedivity in electrical impedance tomography of cryosurgery. IEEE Trans. Biomed. Eng. **52(4)**, 695 (2005)
3. T. K. Bera , Bioelectrical Impedance Methods for Noninvasive Health Monitoring: A Review. Journal of Medical Engineering **2014**, 381251, 28 (2014)
4. Analog Devices Inc. http://www.analog.com/media/en/technical-documentation/data-sheets/AD9764.pdf

Inaccuracy of acoustic measurements in dual-frequence method of sounding

Mariya Kostina*, *Yulia* Shulgina, and *Alena* Chudinova

Tomsk Polytechnic University, 634050 Tomsk, Russia

Abstract. The article describes the new method of defining time coordinate of the moment when echo-impulse comes. The principle of the method consists in consecutive sounding of the (head) well with the signals of different frequencies and analysing the received reflected signals. The article shows the graphs with dependence of measurement inaccuracy when the operating threshold of the comparator for different ratios of frequencies changes. If the comparator works out in different (in order) periods of the received signals then measurement inaccuracy will be 1-2 periods of the carrier frequency.

1 Introduction

Echo acoustic method gained widespread use in modern industry thanks to its advantages like decent accuracy limited by the radiator wavelength, non-contact capability, the ability to measure geometric parameters of the objects and measure a great range of distances. The peculiarity of the method is the possible loss of accuracy in the calibration of distance because of the incorrect defining of the moment when the reflected impulse comes [1-3].

While passing through the acoustic path the signal distorts and handover of the rising flank of the envelope takes place because of time delay in spreading modes of different kinds.

The vital task is the development of the universal method for the analysis of acoustic impulse that enables to gain the information about the depth or distance to the object with the proper accuracy.

Application of comparatively low frequencies in acoustic location leads to the propagation time and the period of sound oscillations to become commensurable quantities and that is why, as a rule, the instantaneously received radio impulse and not its envelope is fed on the input of the comparator. If to use a square-wave modulation or a similar modulation form maximum inaccuracy will be equal to the quarter period of the carrier frequency [4, 5]. The regulating system AGC (automatic gain control) is used to increase the accuracy of acoustic measurements but at the same time, there exist the tasks with the solutions that cannot be essentially resulted in the signal with ramp amplitude of the rising flank of the echo-signal envelope. The task of defining the time position of the echo-impulse is vital for acoustic devices with the acoustic path, which is the essential

* Corresponding author: mariyakostina91@mail.ru

contribution into changing the shape of the signal envelope. The devices of this type can be distance finders and depth finders [6-8].

The basic inaccuracy of measuring for ultrasound depth finders and distance finders is associated with inaccuracy of defining the coming of the ultrasound impulse. The use of Hilbert transform to get the echo-signal envelope and further defining of its homepoint requires much calculating which does not take place in hand-held devices [9].

One of the innovative methods for defining the temporary propagation delay of the ultrasound is the search method of zeros of a signal [10].

The authors offer to use the dual-frequence method of sounding to define the distance to the monitored object. The use of the signals with two different frequencies enables to exclude the essential amount of measurement errors[11, 12].

2 Description of dual-frequence method of sounding

The principle of the new method is radiating of two signals at different frequencies and measuring two time intervals between the radiated and the received signal according to the moment when the signal reaches the triggering value (Figure 1) [13, 14].

In the result we have two time coordinates t_1 and t_2 and the difference between them will depend on sounding frequencies and the number of period when the comparator has worked out [15]. Calculation for the time position of the received echo-impulse is done in accordance with the time intervals t_1 and t_2. Comparison of these time intervals and their correction takes place after calculating the time intervals between the radiated and the received signals and is done according to this expression:

$$(\Delta t_1 - i \cdot T_1) - (\Delta t_2 - i \cdot T_2) = \min , \tag{1}$$

where T_1 – the period of oscillations of the first ultrasound wave, T_2 – the period of oscillations of the second ultrasound wave, i– the number of correction, Δt_1– the first measured time interval, Δt_2– the second measured time interval. The expression $(\Delta t_1 - i \cdot T_1)$ is used to define the distance to the reflective surface.

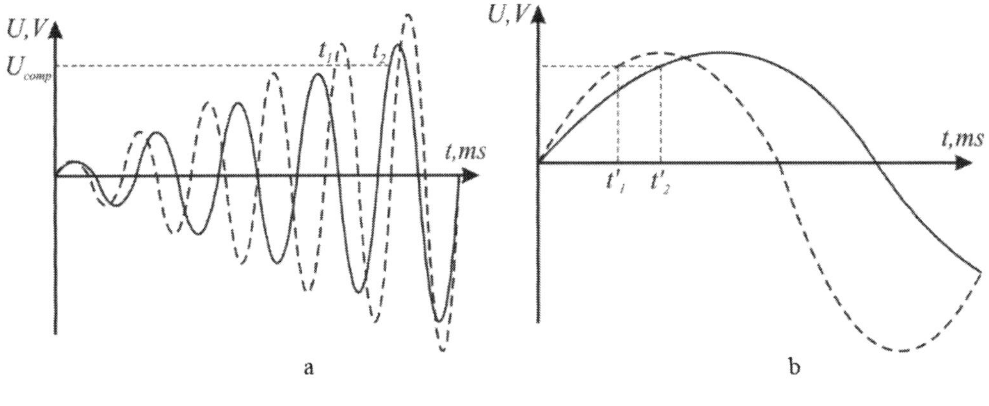

Fig. 1. Oscillograms of the initial part of the two echo-signals (full line graph - the first echo-signal with the repetition period T_1) a - the moment when the comparator has worked out, b - the result of performing the correction where U_{th} – threshold voltage of the comparator; t_1, t_2 – the time when the comparator has worked out, for the frequencies 1 and 2 respectively; t'_1, t'_2 – time intervals after performing the correction.

3 Inaccuracies of the method

3.1 Basic inaccuracy of the method

Measuring inaccuracy of the method is determined by the phase when the comparator has worked out. If the comparator has worked out for the signals with two different frequencies in the same (in order) period of the signal from the moment of its appearance then inaccuracy will be in the range 0 – T/2. To reduce inaccuracy one should increase the frequency of the radiating signal, however, it decreases the range of the measuring distances. The other variant is implementing the phase correction for data processing. Defining the phase of the signal when the comparator has worked out is possible if to analyze the signal at the output of the comparator. Figure 2 explains the principle of calculating the signal phase that takes part in the correction of the calculated distance [16-19].

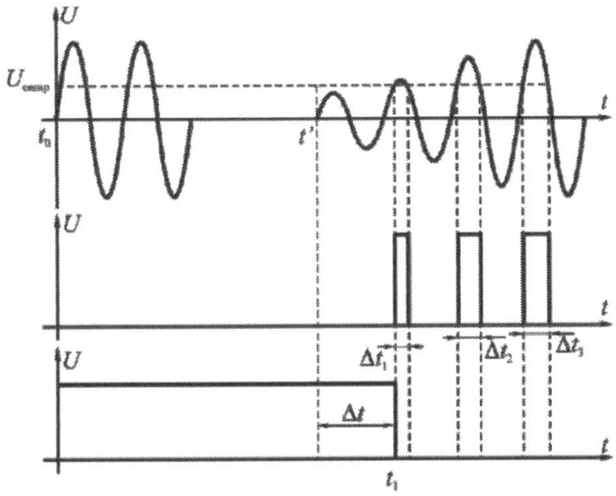

Fig.2. Phase correction of the measurement result based on the signal analysis on the output of the comparator.

Two actions enable to reach high accuracy without sufficient influence on the operation time for the calculation: defining the time coordinate for the point of the first period of the received signal and phase correction of the found time coordinate [20].

Propagation time for using phase correction will be calculated according to the formula:

$$t_0 = (\Delta t_1 - i \cdot T_1) - \varphi \tag{2}$$

where φ – correctional coefficient which is proportional to the phase of the signal at the moment when the comparator has worked out:

$$\varphi = \frac{T/2 - \Delta t}{2} \tag{3}$$

3.2 Response of the comparator in the different periods when processing the signals of different frequencies

The worst errors in measurements appear in cases when the comparator works out in different (in order) periods of the signals with different frequencies. It happens because of different attenuation coefficient for different transmission frequencies. In case when the radiating frequencies differ from each other by several times it is more possible for this inaccuracy to appear. The error in this case will be 1–2 periods because defining the first period of appearing is impossible while performing the iterations [21, 22].

To define the possible measurement errors the series of experiments was carried out. It is necessary for the voltage level at which recording of the time interval happens to be higher than noises for the false responses of the device to be avoided that is why the level of the triggering value lower than 0,5 B is not considered in the experiment [23]. At the same time it is essential for the fixed triggering value of voltages not to be higher than the maximal level of the signal. The response does not occur in other circumstances. Figure 3 shows the graphs of the received signals for different ratios of the frequencies and their corresponding graphs of the measurement errors [24].

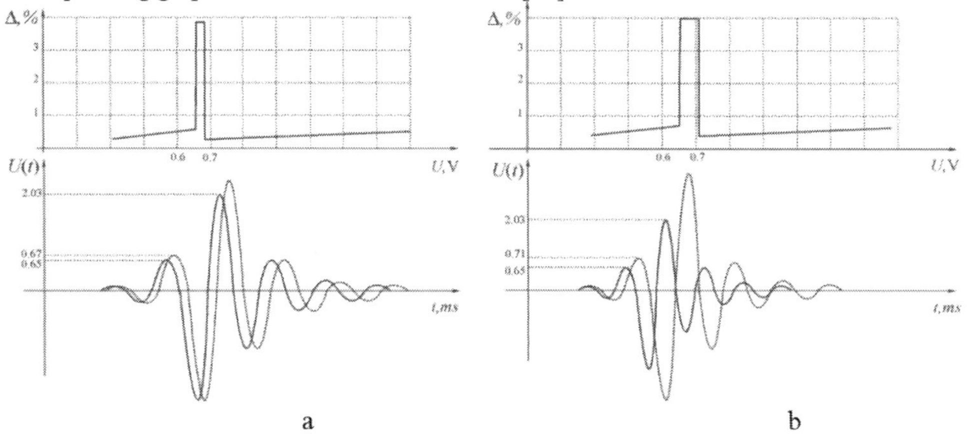

Fig. 3. Dependences of measurement errors when the response level changes from 50 mV to 2 V, for the case a) f1=950 Hz, f2=1000 Hz, b) f1=800 Hz, f2=1000 Hz.

The graphs shown in Figure 3 enable to give the right triggering value of the response of the comparator avoiding the essential measurement values and to make changes in processing the signal. In the first time intervals of signal growing, at the moment of signal appearance in the receiver its level is comparable with the level of noises [25-27]. That is why big errors are possible when setting the triggering value of voltage on the level of the first period of the received signal. In this case, the waveguide is often noisy and false responses of the comparator are possible. When the comparator works out in the same period (in order) in relation to its appearance, we have the measurement inaccuracy that does not exceed 1% from the measured depth for the both frequencies.

Maximal inaccuracy is in the range 3–4% and appears in the case when response of the comparator happens at different (in order) periods of signals for the chosen frequencies. The possibility of the comparator to work out in different (in order) periods increases with the growth of the difference between the frequencies of the signals. The reason is that the signals of higher frequency attenuate faster and, consequently, come to the receiver of the signal with the lower amplitude [28].

To minify the measurement error we implement the signal processor – the module of automatic gain control (AGC) that enables to adjust the signals to amplitude, reduce the possibility of the comparator to work out in different periods.

To prevent the possibility of the comparator to work out in different periods it is also necessary to do 3–5 consecutive measurements. In this case, one can change the triggering value of the voltage to get the stable result [29, 30].

4 Conclusion

The described dual-frequency method of sounding used with the properly chosen threshold voltage and the ratio of frequencies gives inaccuracy less than 1% from the measured depth.

In case when the comparator works out in different (in order) periods of the received signals the measurement error will be 1-2 periods of the carrier frequency. If this happens, AGC systems help to avoid the big error.

Implementation of phase correction into the processing of the received signals helps to increase the accuracy of acoustic measurements when applying the dual-frequence method of sounding.

References

1. L. Huageng, F. Xingjie, E. Bugra, AIAA Journal, **47**, 923-932 (2009)
2. B. Basu, M. Basu, EPE, 1–7. (2011)
3. L. Mažeika, L. Draudvilienė, Ultrasound, **64** (2009)
4. A. Molinaro, Y. Sergeyev, IMEKO, **30**, 187–196. (2001)
5. D. Grimaldi, IEEE TIM, **55**, 5–13 (2006)
6. L. Mažeika, V. Samaitis, K. Burnham, K .Makaya, ULTRAGARSAS, 2011, **66**, 7-12 (2011)
7. L. Mažeika, L. Draudvilienė, ULTRAGARSAS, **65**, 7-12 (2010)
8. L. Angrisani, , A.Baccigalupi, R. SchianoLoMoriello, IEEE TIM, **55**, 442–448 (2006)
9. F. E. Gueuning, M.Varlan, , C. E. Eugène, P. Dupuis, IEEE TIM, **46**, 1236–1240 (1997)
10. K. Huang, Y. Huang, Sensors and Actuators A: Physical, **149**, 42–50 (2009)
11. S. S.Huang, C. F. Huang, K. N. Huang, M. S. Young, RSI, **73**, 3671 (2002)
12. L. Angrisani, A. Baccigalupi, R. S. Lo Moriello, IEEE TIM, **55**(4), 1077–1084 (2006)
13. Y.V. Shulgina et al., *SIBCON*, 7998539 (2017)
14. M.A. Kostina et al., *SIBCON*, 7998536 (2017)
15. A.I. Soldatov et al., *SIBCON*, 7491870 (2016)
16. A.A. Soldatov et al., *SIBCON*, 7147305 (2015)
17. Y.V. Shulgina et al., S*IBCON*, 7491815 (2016)
18. A.I. Soldatov et al., IOP Conf. Ser.: Mater. Sci. Eng., 012117 (2015)
19. A.A. Soldatov et al., *SIBCON*, 7147308 (2015)
20. Y.V. Shulgina et al., IOP Conf. Ser.: Mater. Sci. Eng., 012103 (2015)
21. E.M. Shulgin et al., MEACS, 6986888 (2014)
22. J.V Chiglintseva et al., *SIBCON*, 5044876 (2009)
23. M.A. Kostina et al., *MEACS*, 7414918 (2015)
24. M.A. Kostina et al., Journal of Physics: Conference Series, 012054 (2016)
25. A.A. Soldatov et al., Russ J of NONDESTRUCT, **48**, 255-258 (2012)
26. A.A. Soldatov et al., Russ J of NONDESTRUCT, **48**, 268-271 (2012)
27. P.V. Sorokin et al., NDT, 466-472 (2013)
28. A.I. Soldatov et al., Russ J of NONDESTRUCT, **49**, 625-630 (2013)
29. A.I. Soldatov et al., Russ J of NONDESTRUCT, **49**, 631-635 (2013)
30. A.I. Soldatov et al., *MTT*, 4493172 (2005)

Computer-aided design system for control moment gyroscope

Tamara Kostyuchenko*, and *Nelya* Indygasheva

Tomsk Polytechnic University, 634050 Tomsk, Russia

Abstract. Control moment gyroscope or two-axis balking gyroscope is used in spacecraft orientation systems. The control moment gyroscope is a complex electromechanical system with different parameters which connect by means of specific instant and inverse dependences. This fact could complicate the design process as of getting the optimal parameters the search of construction variations becomes harder. To simplify the control moment gyroscope designing process and increase its efficiency the computer-aided design system could be as highly useful. The computer-aided design system which is used in the present research and described in the article consists of five main subsystems. They cover the main stages of control moment gyroscope designing.

1 Introduction

Modern spacecraft orientation and stabilization systems are based on using the final controlling elements [1]. These elements create the command torque in the stabilization and programmed turns regimes of spacecraft in relation to directional reference. Control moment gyroscope (CMG) is one of such elements. The CMG is a two-axis balking gyroscope which is used for the precision stabilization and orientation of spacecraft. The CMG maintains the required orientation during the flight and prevents the random rotation of spacecraft.

Computer-aided design system (CAD-system) is an automated computer system which provides the advanced facilities for designing. CAD-system is an organizational and technical system for realization the automation designing process and consists of a complex technical, software and other means [2-8].

CAD-system in application of designing the CMG allows automating the designing process. It makes possible to decrease the designing time and improve the final quality of the device.

2 Methods

Currently, there is a large number of CAD-systems which allow creating the 3D models of construction details.

* Corresponding author: ktg@tpu.ru

These 3D models are the basis as for the design documentation and production as for the further simulation in order to test the device characteristics.

The russian software complex T-Flex is one of such type of CAD-systems. It contains the wide range of necessary tools and means for designing. The main advantage of the T-Flex system is the parametric capabilities which allows optimizing the developing construction during the designing process.

The block diagram which is shown in the Figure 1 presents the structure of the CMG designing process in T-Flex system. The whole designing process is distributed between five subsystems which are deal with the certain design calculations. Thus, the results from each subsystem will be accumulated to perform the final parameters of the developing CMG. These subsystems are the following:

- geometric modeling subsystem;
- subsystem of operational characteristics calculation;
- subsystem of strength characteristics calculation;
- design documentation subsystem;
- subsystem of technological preparation of production.

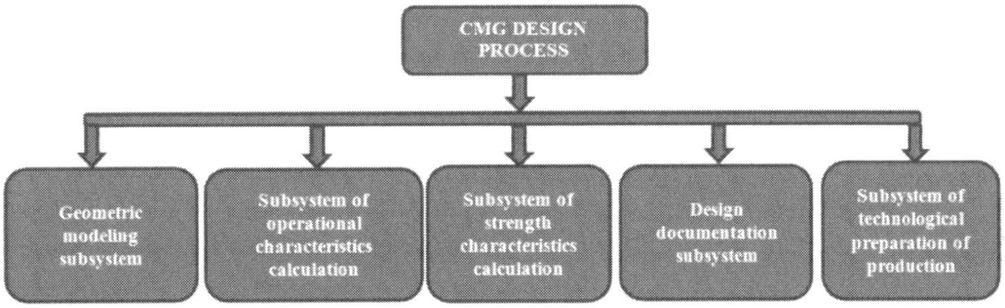

Fig. 1. Block diagram of the CMG design process.

2.1 Geometrical modeling subsystem

The geometric modeling subsystem is meant to create the 3D model of CMG. In particular, it is possible to create the 3D models of construction details, assemblies and subassemblies. The Figure 2 presents the main operations which are provided in the geometric modeling subsystem of T-Flex.

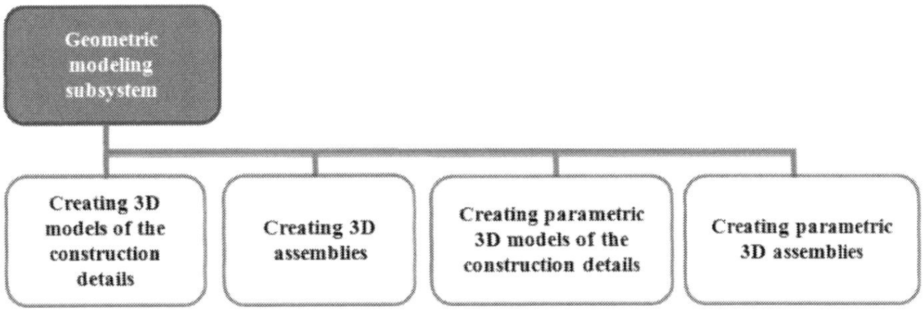

Fig. 2. Block diagram of the CMG designing in geometric modeling subsystem.

2.2 Subsystem of operational characteristics calculation

This subsystem is meant to calculate the required operational characteristics and select the optimal construction variant by the specific criteria. The operational characteristics include weight-size parameters, critical angular velocity, resisting moment, operational life and others.

T-Flex has the advanced capabilities which allow realizing the operational characteristics calculation depending on the geometry. Using the parametric advantages new operational characteristics can be calculated upon the changing of the geometry. After multiple changes of geometry (search the construction variations) the optimal design will be found according to the required criteria.

The subsystem of operational characteristics calculation is based on the use of the T-Flex CAD 2D/3D module. The block diagram of this subsystem is shown in the Figure 3.

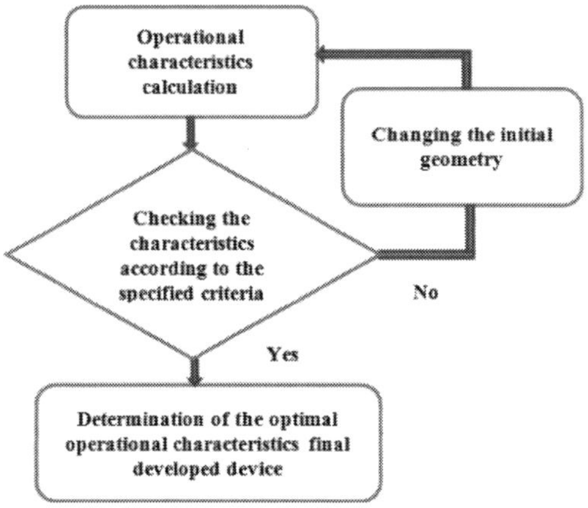

Fig. 3. Block diagram of the subsystem of operational characteristics calculation.

2.3 Subsystem of strength characteristics calculation

This subsystem is meant to strength characteristics calculation. The subsystem is based on T-Flex Analysis module.

Module T-Flex Analysis is based on the using of the finite element method (FEM). It allows providing a range of various simulations: mechanical strength, tuned frequencies, stability analysis, thermal analysis and others. This case, the associative relations between geometric 3D model and finite element model are maintained. The grid finite element model will be reorganized after the changes of the initial 3D model. This process is shown in the Figure 4.

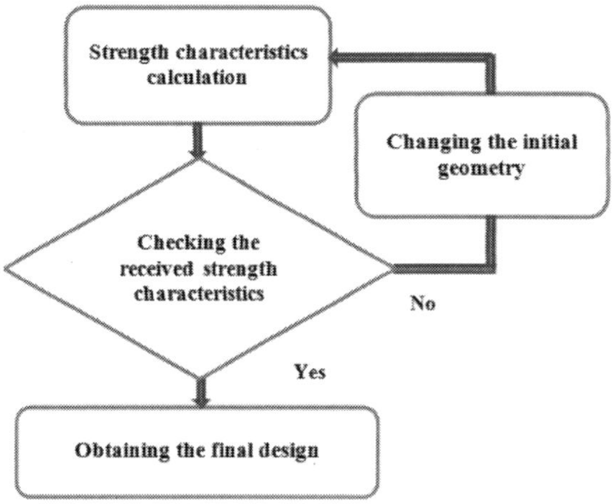

Fig. 4. Block diagram of the subsystem of strength characteristics calculation.

2.4 Design documentation subsystem

The design documentation subsystem is based on the use of the T-Flex CAD 2D/3D module. This module allows creating the design documentation in according to the State Standards requirements. The Figure 5 shows the block diagram of the design documentation subsystem.

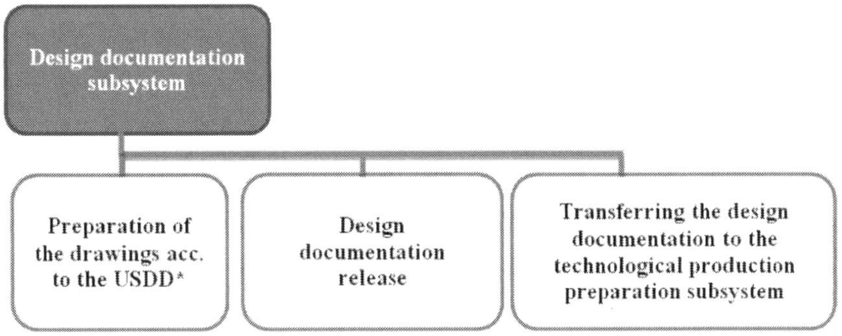

Fig. 5. Block diagram of the design documentation subsystem: * - Unified System for Design Documentation.

At this stage, the design documentation is prepared. Then, drawings and 3D models of the construction details are transferred to the technological preparation of production.

2.5 Subsystem of technological preparation of production

The technological production preparation is a set of connected processes which are ensure the technological readiness to produce the products with the required quality, following up to the special terms, quantity and costs. The necessary operations and stages and their content depend on the production type, construction and products application. Under the technological readiness could be understood the availability of a complete set of technological documentation and technological equipment which are necessary for the production [9].

Technological preparation should be carried out in accordance with the requirements of the Unified System of Technological Preparation of Production [10].

The Figure 6 shows the block diagram of the subsystem of technological preparation of production.

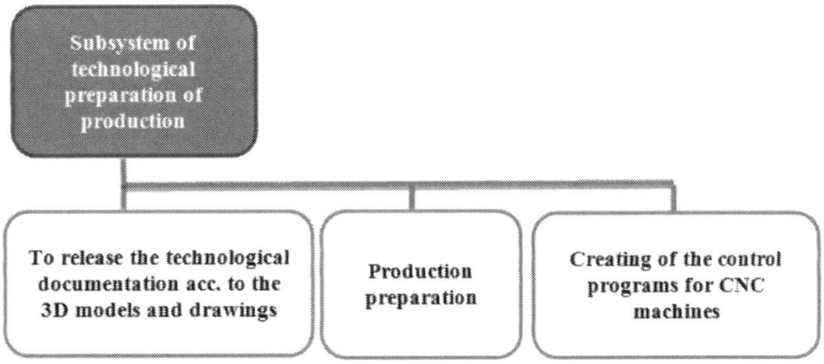

Fig. 6. Block diagram of the subsystem of technological preparation of production.

3 Conclusion

The CAD-system in application of designing the CMG allows developing the complex electromechanical device fast and with the high quality. At the same time, the optimal operational characteristics can be obtained. All these aspects are the basis for the device production.

References

1. P. Baranov et. al. *14th IMEKO TC10 Workshop on Technical Diagnostics,* 90, (2016)
2. K. Lalit Narayan, *Computer Aided Design and Manufacturing* (EDP Prentice Hall of India, New Delhi, 2008)
3. Michael Phiri, *Information Technology in Construction Design* (EDP Thomas Telford Publishing, London, 1999)
4. A. V. Taracenko, V. S. Dmitriev, T. G.Kostuchenko, *KORUS,* 83, (2005)
5. Yu. Britova, V. Dmitriev, T. Kostyuchenko, IOP Conference Series: Materials Science and Engineering **132(1)**, (2016)
6. V. Dmitriev, Y. Britova, MATEC Web of Conferences 01005, (2016)
7. V. Ogay, P. Baranov, A. Stepankova, IOP Conference Series: Materials Science and Engineering **66(6)**, (2014)
8. A. G. Dolgih, V. M. Martemyanov, *SIBCON,* 7491844 (2016), DOI: 10.1109/SIBCON.2016.7491844
9. D. A. Madsen, *Engineering Drawing and Design* (EDP Clifton Park, NY, 2012)
10. G. Farin, J. Hoschek, M. Kim, *Handbook of computer aided geometric design* (EDP Elsevier, 2002)

Calibration setup with metrological support for space qualified electric field probe

Alexey Kozlov[1,*], *Alexander* Shilov[1], *Alexey* Styuf[1], *Alexander* Doroshkin[1], and *Anatoly* Nikitenko[1,2]

[1]Novosibirsk State University, 630090 Novosibirsk, Russia
[2]Siberian State University of Water Transport, 630099 Novosibirsk, Russia

Abstract. The article shows the results of calculations necessary for testing to approve the space qualified electric field probe used in outer space. A special electric field probe calibration setup was developed and manufactured, which allows testing a space qualified electric field probe in the entire measurement range (from ± 0.1 to ± 200 kV/m) under low (-80 °C) and high (+72 °C) temperatures and at reduced pressure of $1.3 \cdot 10^{-3}$ Pa (10^{-5} torr). Metrological tests of the probes batch, calibrated by the space qualified electric field probe calibrator, performed at the verification equipment under normal conditions, showed that the maximum error of the electric field probe does not exceed 15% in the entire measurements range which will increase the accuracy of determining the satellite surface charging parameters and help in designing modern spacecrafts.

1 Introduction

Charging the satellite surface due to low-energy particles flows has been studied since the early 1970s [1] and up to the present [2-4]. This phenomenon plays an important role for satellites located in highly elliptical and polar orbits, where the intersection with the Earth's radiation belts is regular [5]. The spacecraft passing through the magnetosphere with a high concentration of charged particles leads to the charge accumulation on the dielectric surfaces and induce surface electrostatic discharges, which have a destabilizing effect on the onboard equipment. Therefore, it is important to monitor the charge levels on the spacecraft surfaces.

Also, this problem is actual for small spacecrafts [6, 7], in particular, of the CubeSat type [8], since a significant part of such satellites locate in solar-synchronous and polar orbits, with a frequent crossing of the Earth's radiation belts. Today the massive use of nanosatellites for educational purposes stimulates the development of scientific programs using small spacecrafts, and a lot of commercial programs have been launched in this connection. In these conditions, the spacecraft resource becomes important, and this means that destabilizing factors and, in particular, the spacecraft surface charging should be controlled.

* Corresponding author: Alexey.koslov@yandex.ru

The desire to expand the functionality of small spacecrafts stimulates the use of different jet propulsion devices. When accelerated charged particles are the basis of such an engine [9, 10], charge accumulation can take place and, if it is not compensated under control, the engine thrust goes to zero. Thus, monitoring the spacecraft charge level is also very important in this case.

Controlled charging of dielectric surfaces is also needed in laboratory studies of the materials charging by plasma direct impact [11, 12], as well as in microcircuits and materials radiation tests [13, 14].

Novosibirsk State University has developed several generations of diagnostic equipment [for example, 15, 16], covering the above-mentioned issues of surface charging control. To measure the spacecraft surface charge, the space qualified electric field probe is used [17, 18]. Probe's technical specifications: dimensions, mass, power consumption and temperature range, Figure 1, allow applying it in CubeSat satellites.

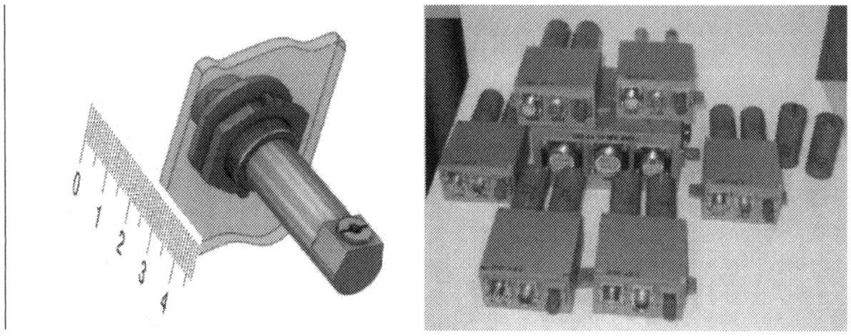

Fig. 1. Electric field probe (left) and surface charge control equipment (right).

A thorough study of surface charging processes requires to use measuring instruments with approved metrological characteristics. However, the existing secondary standards (for example, the verification equipment for measuring the electric field intensity P1-23 [19]) do not allow testing electric field probes in conditions close to the operation modes on the spacecraft. Due to the fact that the probe is designed taking into account operation in outer space, its operability, including metrological characteristics, should be approved throughout the temperature range and under reduced pressure. However, today it is possible to determine its metrological characteristics only in normal climatic conditions, since the Federal Information Fund for Ensuring the Uniformity of Measurements lacks the measuring standards necessary to determine the probe metrological characteristics under conditions identical to its operating conditions.

This article is devoted to the metrological certification of the standard, which will be used to approve the space qualified electric field probes in conditions close to the outer space environment.

2 Calibrator description

In order to solve this problem, the KDEP-200 electric field probe calibration setup was designed as a flat capacitor, which was adapted to be installed in a thermal chamber or thermovacuum chamber to test electric filed probes in the entire temperature range (from -80 to +100 °C) and at reduced pressure from $1.3 \cdot 10^{-4}$ to $1.3 \cdot 10^{-2}$ Pa (from 10^{-6} to 10^{-4} torr). The general view of the calibrator is shown in Fig. 2.

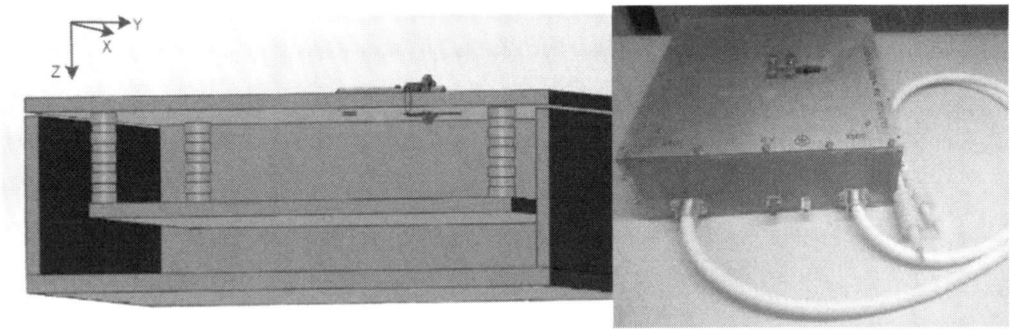

Fig. 2. KDEP-200 electric field probe calibrator (left); General view of KDEP-200 (right).

Based on the results of the finite elements method, it was found that the field strength in the probe measurement area differs from the field strength in the flat capacitor by no more than 0.16% (see Figure 3).

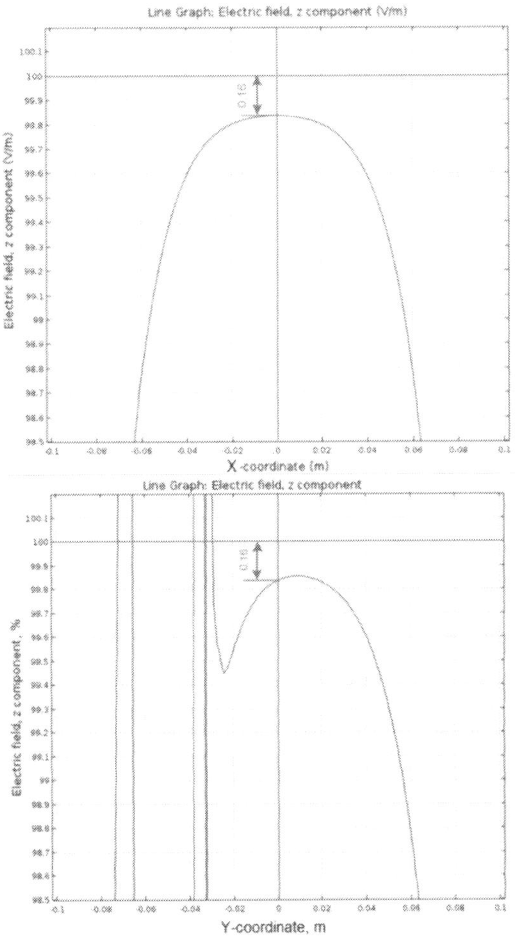

Fig. 3. Distribution of the Z-component of the electric field strength along the coordinate X (left) and the coordinate Y (right).

The final error of the calibrator is also affected by the capacitor thermal linear expansion, change in the dielectric permittivity of the medium under various conditions of

use, and metrological characteristics of additional measuring instruments (high voltage divider, voltmeter, slide caliper). The resulting relative extended measurement uncertainty at KDEP-200 with a confidence probability P = 95% does not exceed 5%, which corresponds to the maximum permissible standard accuracy intended for probes calibration having an accuracy of 15%. Based on the results of the KDEP-200 certification performed in 2017, the relative extended uncertainty of measurements on the standard was 2.28%.
In order to compare the KDEP-200 calibrator with the secondary standard, similar to the state primary standard of the electric field strength unit, the first calibration of the qualified electric field probes batch was performed at KDEP-200. The tests of this batch under normal conditions at the P1-23 reference equipment showed that the maximum deviation of the field strength measured by the probe does not exceed 15% of the P1-23 readings (Figure 4), which meets the requirements for the electric field probes. Further tests in the entire temperature range, including tests under reduced pressure, and determination of the electric field probes metrological characteristics can also be performed.

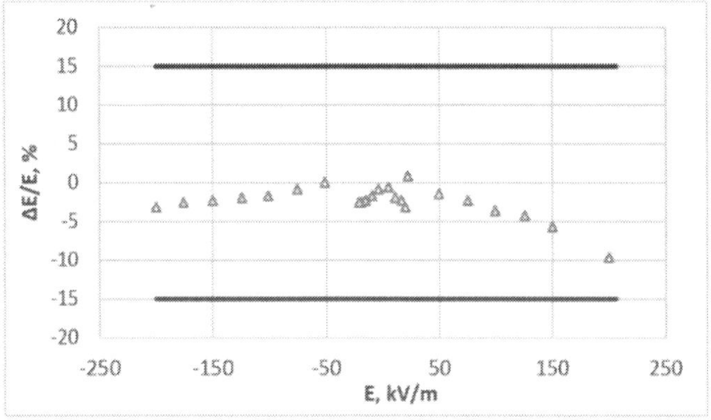

Fig. 4. Deviation of the calibrated by KDEP-200 electric field probe's readings from the P1-23 readings.

3 Conclusion

Thus, the developed KDEP-200 calibration setup meets the metrological requirements for standards and will be applied to calibrate the space qualified electric field probes. This will increase the accuracy of determining the spacecraft's surface charging parameters and can be used to improve charging models in modern spacecraft's design, including small CubeSat satellites. This achievement will allow controlling the charging destabilizing effect on onboard equipment, optimizing the operation of small engines based on accelerated charged particles, and providing reliable determining the material static characteristic in laboratory studies using plasma methods and radioactive tests.

This work was supported by the Ministry of Education and Science of the Russian Federation: Project No. 14.575.21.0154, Project Identification No. RFMEFI57517X0154.

References

1. S. E. DeForest, J. Geophys. Res., **77**, Feb. 1, 65 (1972)
2. X. Meng, D Chen, L. Shi, S. Liu, S. Chen, IEEE Transactions on Plasma Science **45(8)**, 2013 (2017)

3. I. Sillanpää, N.Y. Ganushkina, S. Dubyagin, J.V. Rodriguez, Space Weather **15 (12)**, 1602 (2017)
4. Y. Xiangqian, C. Hongfei, Z. Qiugang, W. Jianzhao, S. Weihong, Z. Hong, Z. Jiqing, Z. Weiying, C. Zhe, S. Sipei, J. Xianghong, IEEE Transactions on Nuclear Science **64(11)**, 2822 (2017)
5. V. V. Ivanov, I. A. Maksimov, N.N. Sitnikova, V. V. Khartov, Yu. M. Prokop'ev, Kosmonavtika i raketostroenie [Cosmonautics and rocket production] **1(30)**, 102 (2003)
6. D. Bock, P. Laufer, and M. Tajmar, 30 th International Vacuum Nanoelectronics Conference (IVNC), 1560 (2017)
7. R.M. Albarran, A. Barjatya,A. Journal of Spacecraft and Rockets **53(3)**, 393 (2016)
8. ISO 17770:2017(E) «Space systems -- Cube satellites (CubeSats)»
9. Fernando Mier-Hicks Thesis: Ph. D., Massachusetts Institute of Technology, Department of Aeronautics and Astronautics, 2017.
10. D. Bock, P. Laufer, M. Tajmar, 2017 30th International Vacuum Nanoelectronics Conference, IVNC 2017, 96 (2017)
11. A.A. Chirov, A.B. Nadiradze, V.V. Shaposhnikov Journal of Surface Investigation: X-Ray, Synchrotron and Neutron Techniques **8(4)**, 689 (2014)
12. K. Chou, and J. Wang, Review of Scientific Instruments **88**, (2017).
13. E.I. Rau, A.A. Tatarintsev, E.Y. Zykova, Physics of the Solid State **59(8)**, 1526 (2017)
14. L. K. Sarno-Smith, B. A. Larsen, R. M. Skoug, M. W. Liemohn, A. Breneman, J. R. Wygant, and M. F. Thomsen, Space Weather **14(2)**, 151 (2016)
15. L. I. Aristov, M. S. Borodin, Y. M. Prokopiev, Scientific-technical development of KB "Salyut" for 2009-2011 **3**, 327 (2012)
16. A. Pelemeshko, A. Styuf, V. Prokopyev, Y. Prokopyev, A. Zadorozhny, MATEC Web of Conferences **102**, 1 (2017)
17. Patent RU 2414717 A.M. Shilov, Ju. M. Prokop'ev, V. Ju. Prokop'ev, I.V. Shchepikhin, Priority of 18.01.2010. Registered with the Public Register for Inventions 20.03.2011, Bull. 8
18. A.S. Kozlov, A.M. Shilov, Y.M. Prokopiev, Proc. of V International Forum for Young Scientists «Space Engineering», 7 (2017).
19. The calibration setup for the measurement unit of electrostatic field P1–23. Available at http://www.ciklon-pribor.ru/?catname=%CF1-23 (accessed 31 January 2017)

A new level in a professional training in implementing the educational programs of interactive learning in a digital educational sphere

Alexander Raikov[1], *Kobilzhon* Zoidov[2], *Valery* Loginova[3,*], *Alexander* Chernov [4], and *Vitalia* Bortalevich[5]

[1]RAS, Institute for Control Sciences, 117418 Moscow, Russia
[2]RAS, Laboratory of Integration of the Russian Economy, 117418 Moscow, Russia
[3]RAS, Institute of Market Problems, 117418 Moscow, Russia
[4] RUDN University, 117198 Moscow, Russia
[5]National Institute of Energy Security, 119992 Moscow, Russia

Abstract. The article deals with the complicated system consisted of various aspects of the interactive training which help online operators to reveal and define ways of relation`s development between huge sort of electronic data during the interactive training. For instance, they are training, testing, adjustment of tasks, etc. The electronic content is formed in the automated system of the interactive training realized as a part of digital educational sphere. This system is constructed in order to increase the professional skills of the specialists.

1 Introduction

Designing and implementing successful education programmes of interactive learning is the key to fundamental, wide-ranging educational reforms. The most successful e-learning ventures take place in professional training and professional upgrade programs. These goals can only be achieved if training and professional development programs are well done and carefully developed and implemented.

In the modern world nowadays the interactive learning is one of the most fast-growing spheres of education. Moreover, today the educational programs of interactive learning have a relevant role in the formation and development of the skills and abilities of professionals to interact a multicultural market.

To increase the professional level and skills of a specialist there is need to give the interactive education operators (both individuals and software systems) an opportunity to reveal and analyze various electronic data concerning the learner.

For instance, the most important are his personal skills, background knowledge, cultural level, interests, hobbies, specifics of the character and personal motivation.

* Corresponding author: loginovalerochka@mail.ru

2 Problem statement

It may be realized in two ways: in manual mode or automated mode. The manual mode takes place if the operator is an individual. Then again, if the operator is a software package it is an automated mode.

During the educational process, operators analyze the learner`s behavior while interacting with other environmental factors affecting the student (work, rest, personal life, participation in public, political and social activities, collective network games, intellectual time-spending, creating new things, etc.) [1, 2].

Another key thing to remember is the significant role of self-made work performed by the student during the educational process.

Such revealed connections of cognitive and personal components may have explicitly and implicitly various sorts of information [3].

Modern information technologies have enabled faster, more reliable and comprehensive data collection [4-6]. As part of the educational process, it is also need to create information databases with information found earlier. Such databases can be used as independent sources of information and can be integrated with other similar databases.

Organizing the fast feedback of the results of such reactions is guarantied by the evaluation service of the interactive training system operators.

Educational module means a training package (including tasks, tests, exercises, materials, etc.) adapted to the certain learner which aim is to make the learner reach the successful results in many various spheres.

In future, it will help the worker to make an innovative solution of the complex professional tasks in complicated and stressful conditions.

This interactively realized educational process is based on the solution by the learners the theoretical and practical tasks of a training package. It is dynamically adapted to a certain sphere of the industry [7].

3 Solution

Attributive-semantic analysis helps to analyze the various kinds of tests results on the point how the future worker fits the professional worker model, for instance:

- to state the fact of achieving the certain educational results with an indication of its positive or negative dynamics,
- to determine the structural relationships of cognitive and personality components, changing in the educational process,
- to correct the imprint vector,
- to identify the differences between the programmable model of a industry specialist and the professional model of the worker,
- to reveal the contribution of individual elements of the educational process into building an educational sphere, etc.

This educational process may become a base for supporting new types of cognitive-reflexive imprintation. For this purpose there is need to correct the educational process in accordance with the international standards, needs of the state, market, current (political, economic, social, etc.) situation in the world.

What is more, tests results analysis is one of the most important elements of making higher competent and qualified workers. So the participants of the educational process should be provided with:

- test results of the examinees with operator's opinion on the correspondence of the professional model of the employee;
- evaluating the content of test results for specialist`s compliance of professional model of the employee;

- information on the correlation between the versions of an educational modules and / or its components;
- opinions on tests results of other examinees regarding to the test results presented in the system of interactive learning;
- recommendations and any kind of notes to the tests authors to make the educational process better;
- contribution estimates of the different participants to the whole collective results of the educational process.

Then again, traditional approaches to the formation of a professional model of an employee and his/her personal competence are normally and traditionally based on analysis of examine test results of the learner.

If we use the authors' approach, it is possible to find out the estimates of the interactive training system operators in which the educational module was criticized or, on the contrary, had positive assessments.

However, these connections of cognitive and personal components do not have any sort of information on trainee motivations in contrast to Attributive-semantic analysis of the tests results by the end of the interactive education [8, 9]. Implementation of the authors model allows to optimize process of professional training.

Moreover, while forming a professional model of a worker and his/her personal competences we may use cognitive and personality components of the worker showed during the interactive learning [10-12].

Because of such testing data received helps to analyze thoroughly and evaluate the quality of practical assignments, its influence on workers, to study personalized "portraits" of both individual students and the whole collective. To put it another way, in the end it formulates recommendations and conclusions regarding the profile of personal qualities and professional competences of the worker.

Such steps allow solving a number of new innovative tasks and forming learner`s knowledge package and set of skills. Moreover, the results of such testing can be analyzed not only by the interactive training system operators (both software complexes and individuals), but also by teachers and examiners.

4 Conclusion

The analysis of various interrelations described in this article is one of the key elements of authors` approach to the formation of the profile of personal characteristics and professional competencies of the professional model of the employee.

Another key point to remember is that using of attributive-semantic methods in the system of interactive learning allows formulating a personalized assessment because of the personal competency matrix.

Overall, the conceptual core of the developed educational system is the transition from the standard one to the open, flexible, organizational system of interactive education within the digital educational sphere.

The theses were prepared with the financial support of the Russian State Scientific Foundation (project No. 16-02-00465a "Development of a mechanism for monitoring, modeling and planning of sectoral development in industry in Russia and the EAEC based on the analysis of cooperative dynamics of aggregated economic entities").

References

1. A. A. Soldatov, P. V. Sorokin, A. A. Abouellail, I. I. Obach, V. Y. Bortalevich, Y. A. Shinyakov, M. P. Sukhorukov, *SIBCON*, DOI: 10.1109/SIBCON.2017.7998534 (2017)

2. A. I. Soldatov, S. O. Ogar, *MTT*, 100, DOI: 10.1109/SPCMTT.2005.4493206 (2005)
3. A. S. Bugaev, E. L. Loginov, A. N. Raikov, V. N. Saraev, *S&T. Inf. Pr.* **36(1)**, 68 (2009)
4. A. I. Soldatov, A. A. Soldatov, P. V. Sorokin, E. L. Loginov, M. A. Kostina, O. A. Kozhemyak, S. I. Bortalevich, *SIBCON*, DOI: 10.1109/SIBCON.2016.7491870 (2016)
5. Y. V. Shulgina, A. L. Starostin, M. A. Kostina, T. S. Mylnikova, A. I. Soldatov, *MEACS*, DOI: 10.1109/MEACS.2015.7414918 (2015)
6. A. I. Soldatov, A. A. Soldatov, M. A. Kostina, V. F. Tolkatchev, J. Phys. Conf. Ser., **671(1)**, DOI: 10.1088/1742-6596/671/1/012054 (2016)
7. E. L. Loginov, A. N. Raikov, Therm. Eng. **62(4)**, 233 (2015)
8. A. P. Grigorev, A. I. Soldatov, P. V. Sorokin, *MTT*, 97 (2000)
9. D. N. Demyanovich, O. S. Vadutov, A. I. Soldatov, *MEACS*, (2014)
10. H. Shevrin, W. J. Williams, R. E. Marshall, US Pat. №4699153, (1987)
11. G. Benson, I. Slesarenko, P. Shamritskaya, Adv. I. S&C, **544**, 609 (2017)
12. I. Slesarenko, I. Zabrodina, M. Netesova, Adv. I. S&C, **715**, 252 (2018)

The implementation of internal assessment mechanisms in the management of the educational program: ESG principles and new educational standards in the Russian Federation

Ivan Nikanorov[1,*], and *Antoniy* Shvindt[2]

[1]Committee on Education and Science of the State Duma of the Russian Federation, 125009 Moscow, Russia
[2]Center for Innovative Educational Technologies, Moscow Institute of Physics and Technology (State University), 141701 Moscow, Russia

Abstract. The purpose of this article is to describe possible approaches to transforming the management of the educational program into an educational institution of higher education in the context of implementing the principles laid down in the Standards and Recommendations for Quality Assurance in the European Higher Education Area (ESG), as well as the updated Russian state higher education standards (RSHES 3 ++). The article analyzes the ESG, examines the main models of managing educational programs, as well as their possible transformations in terms of the formation of mechanisms for internal assessment of the quality of educational activities and the training of students, introduced in the updated federal state educational standards.

1 Introduction

The fourth industrial revolution, which led to a gradual transition to a new technological order, significantly changed the economy and the social sphere, placing the synthesis of the digital, physical and biological spheres in the focus of strategic development. All this significantly changed the idea of the disciplinary division of science, because the most end-to-end technologies that have become most in demand one way or another affect a wide range of scientific fields and require a different approach to the organization of science and implementation mechanisms [1]. In a special way this fact updates the new model of the organization of education, since it requires new competencies, high learning speed, an individual trajectory and a flexible schedule that allows to combine training with production activities [2]. In this regard, the classical formalized system of organization of higher education undergoes transformation towards various forms of interaction between participants in educational relations and greater personalization, taking into account both

* Corresponding author: nikanorovin@gmail.com

the needs of students and the need to constantly change professional competencies. This requires a significant transformation of the higher education systems of the CIS and BRICS countries, since they need not only a substantial restructuring of educational institutions, but also more effective integration into the global educational space [3,4].

2 The international context of changes in higher education

The new challenges significantly changed the requirements for both the university and the graduate model, taking personal motivation and the ability to build an independent educational trajectory in a university environment as one of the most important conditions for successful learning. This made it possible to develop a unified approach to a set of requirements throughout the European Higher Education Area, which fixed the student-centered approach as one of the basic principles of higher education. Following the results of the Ninth Conference of Ministers of Education of the European Higher Education Area (including Russia) held in Yerevan on May 14-15, 2015, the Standards and Recommendations for the Quality Assurance of Higher Education in the European Space were approved in a new version and the European approach to quality assurance of joint programs. The key standards in the context of the topic discussed in this article are standards 1.3, 1.6 and 1.9 [5].

In the standard 1.3 of this document the following principle is outlined: Student-centered learning and assessment of achievement Standard: Universities should introduce student-centered learning processes into their programs. The methods by which programs are implemented should stimulate students to take an active role in the joint construction of the educational process.

According to the standard 1.6: Educational resources and the student support system Standard: Educational organizations must ensure that the educational resources and student support services are adequate, accessible and appropriate.

Recommendations for the inclusion of students in the internal assessment of the quality of education are outlined in Standard 1.9: Ongoing monitoring and periodic evaluation of programs. Standard: Universities should monitor and evaluate programs periodically in order to ensure that they reach their goal and meet the needs of students and society. The results of these processes should lead to a continuous improvement of programs. All interested persons should be informed of any planned or undertaken actions with respect to these programs.

Thus, we see that the context of the adoption of the Standards and the recommendations for quality assurance in the European Higher Education Area (ESG) is formed by the ever increasing demands of the society and the learners themselves for the system and the quality of education, the diversification of possible educational trajectories and the development of various forms and ways of obtaining both a formal , and informational education. ESGs presuppose the formation of approaches focused primarily on the student, the search for the most flexible educational trajectories and the recognition of competences acquired outside formal education. The implementation of the above ESG provisions aimed at the student is impossible without the formation of a community of students who are specifically involved in assessing and improving the quality of education, expanding the toolkit of student evaluation, and developing mechanisms for taking this assessment into account both in internal and external evaluation of the quality of education, therefore, the relevance of the development of student bodies (organizations) in the quality of education in the countries of the European Higher Education Area will only increase.

3 Implementation of ESG principles in the Russian education system

The development of state-public interaction in the management of Russian education over the past decades gives grounds to say that a special type of government has been formed: state-public. It is characterized by the fact that subjects implementing policies in the field of education and educational organizations carry out constant interaction in the management education and provision of educational services with subjects representing the interests of society and the population, with their responsible participation in this activity. The task of the state is to provide all the necessary prerequisites for expanding the participation of society in the development of the education system, the formation of a public request for the conditions for the implementation of educational rights, monitoring the implementation of legislation on education, the implementation of state educational standards, the allocation of responsibility for the activities of all participants in legal relations in this sphere.

According to paragraph 29 of article 2 273-FZ The quality of education is understood as a complex characteristic of the educational activity and training of the student, expressing the level of their compliance with Russian federal state educational standards, educational standards, federal state requirements and (or) the needs of the individual or legal entity in whose interests educational activities are carried out, including the degree of achievement of the educational program planned results;

Under the physical person, in whose interests educational activity is carried out in the system of higher education, it is meant to understand the student, and under the legal entity - the employer. Accordingly, the law presupposes the possibility of participation of each of the categories of interest in the process of evaluation and improvement of the quality of education, both external and internal.

Nevertheless, it should be noted that the norms of inclusion of students in the processes of education quality internal assessment, fixed in the formulations of the federal law on education, were not reflected in the existing mechanisms for organizing the educational activity of universities, regulated by subordinate regulatory legal acts, such as, for example, federal state educational standards, and, accordingly, in the procedures for state assessment of the quality of education. Thus, the principles of student-centered instruction and the inclusion of students in monitoring the quality of education laid down in the ESG, as well as in the federal law on education, were not fully implemented in the practice of educational results, because of their consolidation in normative legal acts of the Ministry of Education and Science of the Russian Federation. However, this problem can be resolved in the Russian state higher education standards in the new edition (RSHES 3 ++), developed and approved by now for the majority of training areas, which are enacted from September 1, 2018. In the updated standards the corresponding possibilities are provided in paragraph 4.6.2:

4.6.2. In order to improve the Bachelor's program, the Organization, in carrying out a regular internal assessment of the quality of educational activities and training of students in the bachelor's program, attracts employers and (or) their associations, other entities and (or) individuals, including the teaching staff of the Organization.

Such an approach has already been implemented by educational organizations as an initiative even before its normative consolidation in RSHES [6].

Within the framework of the internal system of assessing the quality of educational activity under the bachelor's program, students are given the opportunity to assess the conditions, content, organization and quality of the educational process in general and individual disciplines (modules) and practices. This provides ample opportunities for the implementation of the student position in various management tools of the university: an effective contract of faculty, university self-examination procedures and others at the discretion of the educational organization.

At the same time, this will significantly change the procedure for state accreditation of educational programs, first proposing a mechanism for evaluating the principles of involving students in assessing the quality of education and implementing a student-centered approach to learning.

4 Model of students' participation in quality assurance of education

Let us consider the model of participation of students in various procedures for guaranteeing the quality of higher education in the context of maximum applicability to the Russian educational environment. It should be noted that the European practice, formed in the framework of the EHEA, examines students from several positions, determined by a various role, function and status [7]:

Student as an information provider:

Feedback is one of the most common ways of involving students in internal quality assurance procedures for higher education. At present, different models are implemented in universities in what way, when and what kind of feedback students provide. The most common way of feedback is after each training course or at least once a semester.

Student as an actor:

Some of the most motivated students are ready to participate as an actor of intra-university quality assurance procedures, and not just act as providers of information. students themselves participate in the development of the questionnaire or do it in close cooperation with the academic and administrative staff of the university. They can be involved in the collection and analysis of feedback data, as well as organizing joint training seminars for teachers and students, as well as offering projects aimed at solving current problems and tasks.

Student as experts:

Given that ESG is subject to quality assurance is the quality of education, not teaching, the role of students in this process is significant. Students themselves can better analyze how teaching helped them achieve their educational results. This allows a number of agencies to guarantee the quality of education to evaluate teaching through the experience of teaching students, based on how it actually helps to effectively organize the learning process. The use of the proposed expert evaluation of students in a specific form involves the application of certain procedures, such as, for example, involving students in working groups and collegiate bodies with the right to vote. The attitude towards students as experts is a consequence of a culture that requires a positive perception from both the academic community and the students themselves.

Currently, the possibility of participation of students as experts in the accreditation of educational programs is substantially limited by the current regulatory and legal framework that defines the requirements for experts and the procedure for examination.

Thus, the adoption of new RSHES will allow to fully implement the principles of internal guarantee of the quality of education in Russia and the participation of students in this process, making them fully responsible for the learning process and the choice of the educational trajectory. The mechanisms of external quality assurance of education in Russia (state accreditation of educational programs) are still archaic, it is likely that it will require a substantial change both in the model itself and in the requirements for experts, including through the inclusion of students and employers in the academic composition of the commission. All this should make the system of higher education in the Russian Federation more dynamic and flexible, allowing to respond to the current challenges facing the educational system.

5 Conclusions

The opportunities, provided by the updated federal state educational standards for higher education, for students to participate in the internal system for assessing the quality of education create the necessary regulatory prerequisites for implementing the academic rights of students enrolled in the management of an educational organization enshrined in the federal law on education, and will open the possibilities for implementing the relevant ESG points in practice of creating and improving educational higher education programs.

References

1. M. Hermann, P. Tobias, B. Otto, *49th Hawaii International Conference* (2016)
2. M. Solovyev, K. Mertins, A. Shagdyrov, MATEC Web of Conferences, 01034 (2017)
3. A. Huisman, A. Smolentseva, I. Froumin, *Palgrave Macmillan*, (2018)
4. I. Froumin, O. Leshukov, *Palgrave Macmillan*, Ch. 8 (2016)
5. Standards and Guidelines for Quality Assurance in the European Higher Education Area (ESG). Brussels, Belgium. (2015)
6. M. Balyasin, L. Carvalho, G. Mihut, *Voprosy obrazovaniya / Educational Studies. Moscow*, **1**, 110 (2016)
7. A. Klepikov, K. Mertins, A. Khasenova, *LXXXII International Research and Practice Conference and II stage of the Championship in Psychology and Educational sciences* (2014)
8. H. Alaniska, E. A. Codina, J. Bohrer et al., *ENQA. Helsinki*, (2006)

Mathematical model of a flexible asymmetrical rotor for active magnetic bearing reaction wheel

Miroslav Polyakov[1,2,**], *Anatoliy* Lipovtsev[1], *Vladimir* Lyanzburg[1]

[1] Stock Company «Scientific & Industrial Centre «Polyus», 634050 Tomsk, Russia
[2] National Research Tomsk Polytechnic University, 634050 Tomsk, Russia

Abstract. The paper introduces the mathematical model of rotor for active magnetic bearing reaction/momentum wheels, used as actuator in spacecraft attitude and orbit control system. Developed model is used for estimation of critical speeds and forced oscillation magnitudes with a glance of the rotor modes. Rotor is considered as a two-mass system, consisting of a shaft and a rim, active magnetic bearings are assumed to be a linear elastic springs. The equations of the rotor motion are derived using the Lagrange equation. Developed model is verified by comparing the calculated Campbell diagrams with the results of the finite-element modal analysis, performed in the ANSYS software.

1 Introduction

Continuous complication of tasks, performed by modern satellites requires the increasing of attitude and orbit control systems operating performance, primarily pointing accuracy and, what is equally important, pointing stability. These parameters are key for high precision optical and astrophysical instruments, used on advanced Earth observation satellites, orbital observatories and telescopes. Nowadays, the need of pointing stability improvement leads to the fact, that the great efforts are applied to solving the microvibration and force-torque disturbances issues. Typically, spacecraft internal disturbances are generated by on-board electromechanical equipment, including reaction wheels (RW) [1-4]. The major RW disturbances sources are ball bearing vibrations, caused by geometrical imperfections, dimensional drift of bearing elements and internal clearances of bearing unit, rotor residual unbalances, determined by the balance quality, control torque ripple, produced by friction torque instabilities [1]. Thus, mainly ball bearings excite the RW micro-disturbances.

Using of active magnetic bearings (AMB) for rotor support is a one of the possible ways to attenuate the RW microvibrations and disturbances. Typically, AMB is a mechatronic system, composed of control object (rotor), actuators (electromagnets), feedback sensors and control system. Significant suppression of the disturbances requires development of the rotor mathematical model, included rotor modes, since they determine the critical speeds and forced oscillations amplitudes inside the gap.

* Corresponding author: polyakovmir91@gmail.com

2 Mathematical model of RW rotor

RW rotor consists of the rotating parts of electric drive and AMB, mounted on the flywheel. To ensure the optimal mass and inertial characteristics of the rotor (the maximal polar moment of inertia J_p with the minimal mass M), flywheel comprises of the massive rim, linked to the shaft by a thin lightweight disk (Figure 1a). Results of the RW rotor modal analysis show, that in the case of rigid supports the first natural frequency generally associates with the angular oscillations of the rim and the disk about the lateral axes (Figure 1b). The value of this frequency is rather low and can be spotted in the operating range of rotational speed [5].

a) b)

Fig. 1. RW rotor (a) and its first mode (b)

This mode shape allows the RW rotor to be seen as a two-mass system, consisting of the shaft and the rim. The shaft is defined by mass M_s, polar and equatorial moments of inertia $J_{p.s}$, $J_{e.s}$, rim is defined by mass M_r, polar and equatorial moments of inertia $J_{p.r}$, $J_{e.r}$. There are the following bounds in the system: linear translations and rotation about spin axis are performed by rim and shaft simultaneously (the bound are rigid). The rim rocking relative to the shaft about transversal axes is described by the angular stiffness R_d.

The following coordinate systems are introduced: XYZ – frame of the RW; $OX_sY_sZ_s$ – rotating frame related to the shaft; $OX_rY_rZ_r$ – rotating frame related to the rim. $Oxyz$, $OX'_sY'_sZ'_s$, $OX'_rY'_rZ'_r$, $OX''_sY''_sZ''_s$, $OX''_rY''_rZ''_r$ – intermediate frames, describe the shaft and rim position after linear and angular displacements (Figure 2).

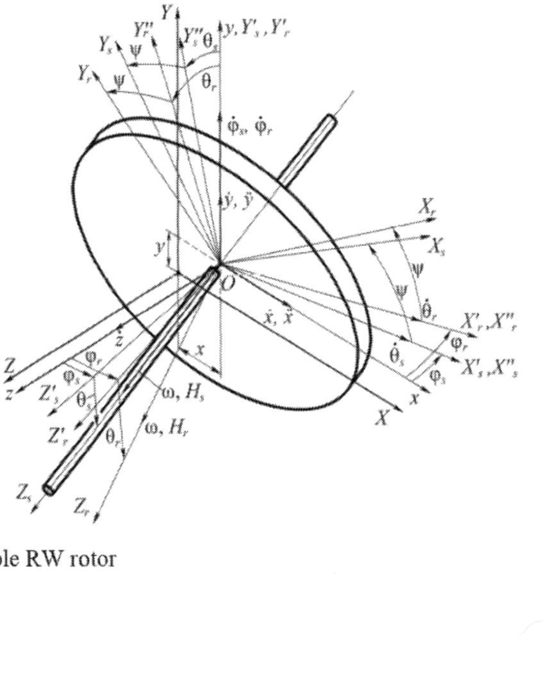

Fig. 2. Model of the flexible RW rotor

Thereby, subjected to restrains, position of the shaft and rim of the RW rotor is describes by eight generalized coordinates: linear translations of the shaft and rim x, y, z, spinning $\psi=\omega t$, shaft and rim rocking about transversal axes with the angles φ_s, θ_s и φ_r, θ_r respectively (Figure 2).

Mathematical model of RW rotor is derived by solving the Lagrange equation for every generalized coordinate q.

Projection of the angular velocity vectors of the shaft and rim $\omega_{s(r)}$ on the axes of the rotating frames $OX_sY_sZ_s$ and $OX_rY_rZ_r$.

$$\omega_{s(r)} = \begin{vmatrix} \dot{\theta}_{s(r)} \cos\omega t + \dot{\varphi}_{s(r)} \cos\theta_{s(r)} \sin\omega t \\ \dot{\varphi}_{s(r)} \cos\theta_{s(r)} \cos\omega t - \dot{\theta}_{s(r)} \sin\omega t \\ \omega - \dot{\varphi}_{s(r)} \sin\theta_{s(r)} \end{vmatrix}. \quad (1)$$

Assuming the rocking angles φ_s, θ_s, φ_r, θ_r are negligible ($\cos\theta_{s(r)}\approx 1$, $\sin\theta_{s(r)}\approx\theta_{s(r)}$) projection of the angular velocity vectors can be written in the following form

$$\omega_{s(r)} = \begin{vmatrix} \dot{\theta}_{s(r)} \cos\omega t + \dot{\varphi}_{s(r)} \sin\omega t \\ \dot{\varphi}_{s(r)} \cos\omega t - \dot{\theta}_{s(r)} \sin\omega t \\ \omega - \dot{\varphi}_{s(r)}\theta_{s(r)} \end{vmatrix}. \quad (2)$$

Axes of rotating frames coincide with the principal axes of inertia, therefore the kinetic energy of the rotor

$$T = \frac{1}{2}\left[M\left(\dot{x}^2 + \dot{y}^2 + \dot{z}^2\right) + J_{e.s}\left(\omega_{sX}^2 + \omega_{sY}^2\right) + J_{p.s}\omega_{sZ}^2 + J_{e.r}\left(\omega_{rX}^2 + \omega_{rY}^2\right) + J_{p.r}\omega_{rZ}^2\right]. \quad (3)$$

Finally, the kinetic energy of the rotor is obtained by substituting equation (2) into equation (3)

$$\begin{gathered} T = \frac{1}{2}\Big\{M\left(\dot{x}^2 + \dot{y}^2 + \dot{z}^2\right) + J_{e.s}\left[\left(\dot{\theta}_s \cos\omega t + \dot{\varphi}_s \sin\omega t\right)^2 + \left(\dot{\varphi}_s \cos\omega t - \dot{\theta}_s \sin\omega t\right)^2\right] + \\ + J_{p.s}\left(\omega - \dot{\varphi}_s\theta_s\right)^2 + J_{e.r}\left[\left(\dot{\theta}_r \cos\omega t + \dot{\varphi}_r \sin\omega t\right)^2 + \left(\dot{\varphi}_r \cos\omega t - \dot{\theta}_r \sin\omega t\right)^2\right] + \\ + J_{p.r}\left(\omega - \dot{\varphi}_r\theta_r\right)^2\Big\}. \end{gathered} \quad (4)$$

The following assumption is made in consideration of potential energy of the system P. Since the objective of this stage is consideration and identifications of the dynamic characteristics of single rotor, so it is sufficient to model the radial and axial AMB in the form of ideal elastic and isotropic springs with stiffnesses C_A, C_B и C_Z. This assumption, firstly, eliminates the effects of control algorithm and electromagnetic processes in AMB and defines the investigated dynamic behavior of the rotor. Secondly, it gives an opportunity to compare analytical results, obtained by solving the developed model, with the results of modal finite-element analysis. Thus, potential energy P is concentrated in the elastic supports C_A, C_B, C_Z and in the elastic link of the rim and shaft R_d, so

$$\begin{gathered} P = \frac{1}{2}\Big[C_A\left(x + a\varphi_s\right)^2 + C_B\left(x - b\varphi_s\right)^2 + C_Z z^2 + C_A\left(y + a\theta_s\right)^2 + C_B\left(y - b\theta_s\right)^2 + \\ + R_d\left(\varphi_r - \varphi_s\right)^2 + R_d\left(\theta_r - \theta_s\right)^2\Big]. \end{gathered} \quad (5)$$

where a, b – distances between center of mass and radial supports, in the case of asymmetrical rotor $a\neq b$.

Derivatives of Lagrangian for generalized coordinates:

$$\frac{d}{dt}\left(\frac{\partial L}{\partial \dot{x}}\right)-\frac{\partial L}{\partial x}=M\ddot{x}+C_A\left(x+a\varphi_s\right)+C_B\left(x-b\varphi_s\right)=M\ddot{x}+Cx+N\varphi_s; \tag{6}$$

$$\frac{d}{dt}\left(\frac{\partial L}{\partial \dot{y}}\right)-\frac{\partial L}{\partial y}=M\ddot{y}+C_A\left(y+a\theta_s\right)+C_B\left(y-b\theta_s\right)=M\ddot{y}+Cy+N\theta_s; \tag{7}$$

$$\frac{d}{dt}\left(\frac{\partial L}{\partial \dot{z}}\right)-\frac{\partial L}{\partial z}=M\ddot{z}+C_Z z; \tag{8}$$

$$\frac{d}{dt}\left(\frac{\partial L}{\partial \dot{\varphi}_s}\right)-\frac{\partial L}{\partial \varphi_s}=J_{e.s}\ddot{\varphi}_s+J_{p.s}\omega\dot{\theta}_s+Nx+\left(R_s+R_d\right)\varphi_s-R_d\varphi_r; \tag{9}$$

$$\frac{d}{dt}\left(\frac{\partial L}{\partial \dot{\varphi}_r}\right)-\frac{\partial L}{\partial \varphi_r}=J_{e.r}\ddot{\varphi}_r+J_{p.r}\omega\dot{\theta}_r+R_d\varphi_r-R_d\varphi_s; \tag{10}$$

$$\frac{d}{dt}\left(\frac{\partial L}{\partial \dot{\theta}_s}\right)-\frac{\partial L}{\partial \theta_s}=J_{e.s}\ddot{\theta}_s-J_{p.s}\omega\dot{\varphi}_s+Ny+\left(R_s+R_d\right)\theta_s-R_d\theta_r; \tag{11}$$

$$\frac{d}{dt}\left(\frac{\partial L}{\partial \dot{\theta}_r}\right)-\frac{\partial L}{\partial \theta_r}=J_{e.r}\ddot{\theta}_r-J_{p.r}\omega\dot{\varphi}_r+R_d\theta_r-R_d\theta_s; \tag{12}$$

$$\frac{d}{dt}\left(\frac{\partial L}{\partial \dot{\psi}}\right)-\frac{\partial L}{\partial \psi}=\left(J_{p.s}+J_{p.r}\right)\ddot{\psi}, \tag{13}$$

where $C = C_A + C_B$; $N = C_A a + C_B b$; $R_s = C_A a^2 + C_B b^2$.

Mathematical model of rotor forced oscillations under forces $F_x(t)$, $F_y(t)$, $F_z(t)$ and torques $M_x(t)$, $M_y(t)$, $M_z(t)$ is in the following form:

$$\begin{cases} M\ddot{x}+Cx+N\varphi_s=F_x(t); \\ M\ddot{y}+Cy+N\theta_s=F_y(t); \\ M\ddot{z}+C_Z z=F_z(t) \\ J_{e.s}\ddot{\varphi}_s+J_{p.s}\omega\dot{\theta}_s+Nx+\left(R_s+R_d\right)\varphi_s-R_d\varphi_r=M_y(t); \\ J_{e.r}\ddot{\varphi}_r+J_{p.r}\omega\dot{\theta}_r+R_d\varphi_r-R_d\varphi_s=M_y(t); \\ J_{e.s}\ddot{\theta}_s-J_{p.s}\omega\dot{\varphi}_s+Ny+\left(R_s+R_d\right)\theta_s-R_d\theta_r=M_x(t); \\ J_{e.r}\ddot{\theta}_r-J_{p.r}\omega\dot{\varphi}_r+R_d\theta_r-R_d\theta_s=M_x(t); \\ \left(J_{p.s}+J_{p.r}\right)\ddot{\psi}=M_z(t). \end{cases} \tag{14}$$

3 Results

Verification of the developed mathematical model of the flexible asymmetrical rotor was carried out by comparing analytical Campbell diagrams, obtained by solving the system (14) with the results of modal finite-element analysis, performed in the ANSYS (Figure 3).

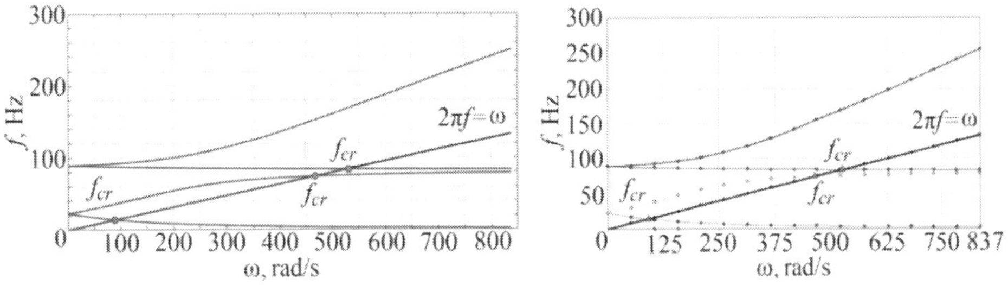

a) b)

Fig. 3. Campbell Diagrams (a – analytical; b – ANSYS)

Excluding the axial translations z and spinning $\psi=\omega t$, an asymmetric system have three critical speeds (Figure 3), on each speed rotor perform a complex motion, combined of translation and rocking, accompanied by natural oscillations. Analytical critical speeds agree with the results of ANSYS modal analysis, relative error less than 1% (Table 1).

Table 1. Critical speeds of rotor in the operating range of rotational speed

Mode	Critical speed, rad/s		Relative error, %
	Analytical	ANSYS	
Rocking, backward whirl	91,1	90,6	0,55
Rocking, forward whirl	469,8	467,1	0,58
Radial translation	533,4	529,8	0,68

4 Conclusions

The developed mathematical model is used for estimation of natural frequencies and critical speeds of flexible rotor of reaction wheel equipped with active magnetic bearings. The relative error of the developed model in comparison with the results of FEM analysis isn't exceed 1%, therefore model can be used for optimization and further investigation of the dynamic behavior of AMB system.

References

1. M. P. Le, *Micro-disturbances in reaction wheels* (Technische Universiteit Eindhoven, 2017)
2. K. Komatsu, H. Uchida, Mechanical Engineering Reviews, **1**, 00079 (2014)
3. R. Hahn, R. Seiler. *14th European Space Mechanisms & Tribology Symposium*, 415 (2011)
4. G. Smet, G. Richardson, S. McLaren, A. Haslehurst. *15th European Space Mechanisms & Tribology Symposium*, 8 (2013)
5. V. Dmitriev, Y. Britova, MATEC Web of Conferences **48**, 01005 (2016)

PADME – new code for modeling of planet georesources formation on heterogeneous computing systems

Viktor Protasov[1,*], *Igor* Kulikov[2], *Igor* Chernykh[2], and *Irek* Gubaydullin[3]

[1]Novosibirsk State Technical University, 630073 Novosibirsk, Russia
[2]Institute of Computational Mathematics and Mathematical Geophysics SB RAS, 630090 Novosibirsk, Russia
[3]Institute of Petrochemistry and Catalysis RAS, 450075 Ufa, Russia

Abstract. Many planets were detected in last few years, but there is no clear understanding of how they are formed. The fairly clear understanding of Solar system formation was founded with time, but there are some doubts yet because we don't know what was at the beginning of the process, and what was acquired afterward. Moreover, formed ideas often couldn't explain some features of other systems. Searching for Earth-like terrestrial planets is another very important problem. Even if any of found exoplanets will be similar to Earth, we couldn't say that it is a "second Earth" exactly because its internal, geological, composition could be different – Venus is a vivid example. A new method for modelling of the planet formation process in a 3D2V formulation based on two-phase approach is presented in the paper. Fluids-in-cells method by Belotserkovskii-Davydov, modified with using the Godunov's scheme, is used to model the gas component. The dust component is described by N-body system solved with the Particle-Mesh method. The method was accelerated by using of Nvidia CUDA technology. Gas-dust disk modelling results with the formation of sealing of gas and dust that could be interpreted as potential exoplanet are given.

1 Introduction

The search of planetary systems is one of the most important tasks of modern astronomy. In the beginning of 90th with the advent of powerful telescopes, many extrasolar planets were discovered. At the end of 2014, its number was about 1800, and at the end of 2017 – more than 3700. The modern views of Solar system formation are based on Kant-Laplace theory, according to which the Sun and the planets were formed at the same time from circumstellar gas and dust. It describes the structure of the Solar system well enough but fails on some other systems. For example, hot Jupiters revolving too close to a star were discovered. Also, questions about the time needed to form the planetary system, and planet formation in multiple systems, are opened [1]. Researching of geophysical features of exoplanets becomes more popular [2] because the understanding of such properties could

* Corresponding author: pro.vik@bk.ru

help not only in searching for Earth-like planets but will allow us to better understand processes inside planets.

There are two main approaches for modelling protoplanetary disks for today – two-component approach [3] and two-phase approach [4, 5]. The second one is used in the paper. To model a gas-component of a disk the Fluids-in-Cells method by Belotserkovskii-Davydov [6], modified with using the Godunov's scheme [7], is applied. Dust particles are described with N-body system which solved by the Particle-Mesh method [8]. To compute the gravitational interaction between gas and particles the Poisson equation for gravitational potential is solved by using the Fast Fourier Transform.

2 Numerical method description

The model of dynamics of a protoplanetary disk in cartesian coordinates is described by the following system of equations:

$$\frac{\partial \rho}{\partial t} + \mathrm{div}(\rho \mathbf{v}) = 0 \tag{1}$$

$$\frac{\partial \rho \mathbf{v}}{\partial t} + div(\mathbf{v}\rho\mathbf{v}) = -grad(p) - \rho grad(\Phi) \tag{2}$$

$$\frac{\partial \rho \varepsilon}{\partial t} + div(\rho\varepsilon\mathbf{v}) = -(\gamma - 1)\rho\varepsilon div(\mathbf{v}) \tag{3}$$

$$\frac{\partial \rho E}{\partial t} + div(\rho E\mathbf{v}) = -div(p\mathbf{v}) - (\rho grad(\Phi), \mathbf{v}) \tag{4}$$

$$div\big(grad(\Phi)\big) = 4\pi G\tilde{\rho} \tag{5}$$

$$p = (\gamma - 1)\rho\varepsilon \tag{6}$$

$$\rho E = \rho\varepsilon + \frac{\rho \mathbf{v}^2}{2} \tag{7}$$

$$\frac{d^2\tilde{\mathbf{x}}}{dt^2} = \frac{\mathbf{F}}{\tilde{m}} \tag{8}$$

$$\mathbf{F} = -grad(\Phi) \tag{9}$$

Equations (1) – (7) describes the dynamics of self-gravitating gas, and equations (8) – (9) – dynamics of any particle. Here ρ – gas density, $\mathbf{v}$ – speed of gas, p – gas pressure, ε – internal energy of the gas, E – full energy of the gas, γ – adiabatic power, Φ – gravitational potential, G – gravitational constant, $\tilde{\rho}$ – density of gas and particles distribution in 3D, $\tilde{\mathbf{x}}$ – position of some particle, $\tilde{m}$ – its mass.

2.1 Gas component modelling

The gas component is modelled using the Fluids-in-Cells method by Belotserkovskii-Davydov. It is based on the scheme of the splitting of physical processes – the solution of initial system of equations is divided into two stages. At the first, Euler, stage the gas dynamics system of equations is solved without advective members – recomputation of parameters is done in assumption that gas is still and has constant density. For a rectangular grid, the approximation is as follows:

$$\frac{\rho_{i,j}^{n+1} - \rho_{i,j}^{n}}{\Delta t} = 0 \tag{10}$$

$$\frac{\rho v_{x_{i,j}}^{n} - \rho v_{x_{i,j}}^{n-1}}{\Delta t} = -\frac{p_{i+1,j}^{*} - p_{i-1,j}^{*}}{h_x} - \rho_{i,j}^{n-1} \frac{\Phi_{i+1,j}^{n} - \Phi_{i-1,j}^{n}}{2h_x} \tag{11}$$

$$\frac{\rho v_{y_{i,j}}^{n} - \rho v_{y_{i,j}}^{n-1}}{\Delta t} = -\frac{p_{i,j+1}^{*} - p_{i,j-1}^{*}}{h_y} - \rho_{i,j}^{n-1} \frac{\Phi_{i,j+1}^{n} - \Phi_{i,j+1}^{n}}{2h_y} \tag{12}$$

$$\frac{p_{i,j}^{n} - p_{i,j}^{n-1}}{\Delta t} = -(\gamma - 1) p_{i,j}^{n-1} \left(\frac{v_{x_{i+1,j}}^{*} - v_{x_{i-1,j}}^{*}}{h_x} + \frac{v_{y_{i,j+1}}^{*} - v_{y_{i,j-1}}^{*}}{h_y} \right) \tag{13}$$

Equations (11) and (12) is obtained from the equation (2) and allows to compute x- and y-components of impulse, respectively. The equation for pressure (13) is obtained from (3) with consideration of (6).

The second stage – Lagrangian. At this stage the advection of the gas is done, all the divergence operators that were not taken into account at the Euler stage are approximated. In simple terms, for each cell, we should compute how much of the gas flowed into it, how much flowed out and calculate the balance:

$$M_{i,j}^{n+1} = M_{i,j}^{n} + \frac{\Delta t}{h_x}(x\_flow\_in_{i,j} - x\_flow\_out_{i,j}) + \frac{\Delta t}{h_y}(y\_flow\_in_{i,j} - y\_flow\_out_{i,j}). \tag{14}$$

To compute the flows through the borders of cells we use the modification of classical approach that considers possible "skew" of the border due to different velocities of the gas in nodes of a cell:

$$x\_flow\_out_{i,j} = \frac{1}{2} \sum_{k=0}^{1} v_{x_{i+1/2,j+(k-1/2)}} \begin{cases} M_{i,j}, v_{x_{i+1/2,j+(k-1/2)}} > 0 \\ M_{i+1,j}, v_{x_{i+1/2,j+(k-1/2)}} \leq 0 \end{cases} \tag{15}$$

2.2 Dust component modelling

The main difficulty in modelling of the particles is in computing of the gravitational force affecting any particles from the others, and the necessity of accounting the interaction between gravitational fields of gas and particles. In the paper, the Particle-Mesh method, which allows reducing the computational costs at this stage of simulation, is used.

The main idea of the method lies in the fact that we compute interaction between particles not as "each-to-each", but projecting them into the similar rectangular grid as used for the gas component and computing the density of particles in each cell with consideration of their masses. Then the Poisson equation for the gravitational potential is solved for that distribution of density. Knowing the potential we could compute the gravitational force easily as $\mathbf{F} = -grad(\Phi)$, where Φ is computed from (5) with consideration that $\tilde{\rho} = \rho_{gas} + \rho_{particles}$.

Such approach allows naturally consider the gravitational interaction between gas and particles, but its accuracy is low. The density of a particle in cell and force affecting it are computed by Clouds-in-Cells method [9] to reduce the error of computations. In this method, the particle coordinates are the coordinates of the center of mass of a "cloud" with finite size. The density of that cloud is distributed among cells into which it fells:

$$\rho_{i,j} = \sum_{c \in clouds} W_{i,j}(x,y) \rho_c(x,y) \tag{16}$$

$$\mathbf{F}_c = -\sum_{i,j} W_{i,j}(x,y) grad(\Phi)_{i,j} \tag{17}$$

$$W_{i,j}(x,y) = \begin{cases} (1 - \frac{|x - x_i|}{h_x})(1 - \frac{|y - y_j|}{h_y}), |x - x_i| \leq h_x \ \&\& \ |y - y_j| \leq h_y \\ 0, else \end{cases} \tag{18}$$

This approach doesn't completely solve the problem, of course, but it significantly reduces the error of computations as shown in [9].

3 Gas-dust disk modelling

Let's consider the region $[-2.5; 2.5] \times [-2.5; 2.5]$, in which is the gas with parameters:

$$\rho(r) = \begin{cases} 2, r < 2 \\ 3, r < 0.25 \end{cases}, \ p(r) = \begin{cases} 1, r < 2 \\ 2, r < 0.25 \end{cases}, \ \omega(r) = 2(2 - r)^2 \qquad (19)$$

Also in region $r \in [0.5; 2.0]$ uniformly distributed 50 000 of particles with mass $m = 5 \cdot 10^{-5}$ and angular velocity $\omega(r) = 2(2 - r)^2 + \varepsilon$, where ε – is a tiny deviation.

The formation of the gas-dust ring is shown in figure 1. The seal consisting mainly of particles is formed inside of it. This sealing could be interpreted as a potential exoplanet.

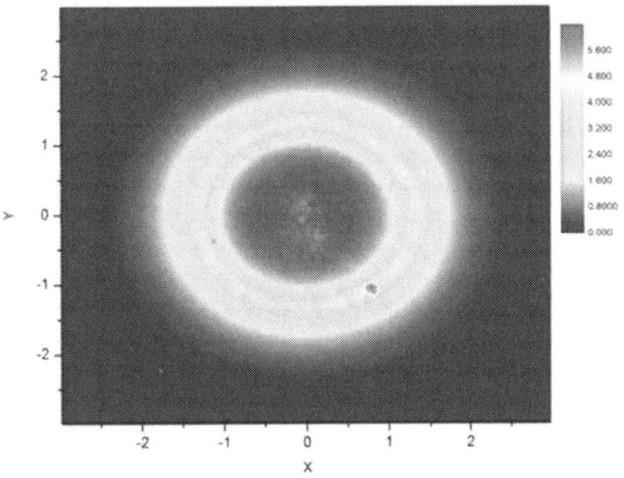

Fig 1. Modeling of gas-dust disk, t = 0.4.

4 Discussion

The described method is not a final point. It is just a timid step on the road to high-quality algorithms of numerical astrophysics which allows answering the most important questions about the Universe highly accurate. At the moment, active work is underway to improve the method. For example, it is possible to sufficiently improve the accuracy of computations if the magnetic field of the gas-dust disk will be taken into account. Yet another way is to recompute the grid accordingly to the medium parameters. In addition, for an adequate modelling of planetary cores, it is needed to consider the interaction between gas and particles not only by gravitational force but by collisions, adhesion, different chemical processes and transition of a substance from one state to another.

5 Conclusions

A new method for modelling of the planet formation process based on the Fluids-in-Cells method and the Particle-Mesh method oriented for using in heterogeneous computing systems is given in the paper. The gas component is modelled by Fluids-in-Cells method by Belotserkovskii-Davydov, modified with using the Godunov's scheme, is used to model the gas component. The dust component is described by N-body system solved with the

Particle-Mesh method. Gravitational interaction between gas and particles is computed by solving the Poisson equation for the gravitational potential with Fast Fourier Transform. Gas-dust disk modelling results with the formation of sealing of gas and dust that could be interpreted as potential exoplanet are given.

The research work was supported by the Grant of the President of Russian Federation for the support of young scientists MK-1445.2017.9, RFBR grants 18-01-00166, 18-07-00757, and 17-31-50023.

References

1. A. Tutukov, A. Fedorova, Astron. Rep. **54**, 305 (2012)
2. C. Dorn et.al., A&A **577**, 83 (2015)
3. S. Inaba et al., A&A **431**, 365 (2005)
4. D. Hubber et al., A&A **529**, (2011)
5. V. Vshivkov et al., Nauchnyj vestnik NGTU **44**, 69 (2011)
6. O. Belotserkovskii, Iu. Davydov, *Metod krupnykh chastits v gazovoi dinamike* (M.: Nauka, 1982)
7. A. Kulikovskii, N. Pogorelov, A. Semenov, *Matematicheskie voprosy chislennogo resheniia giperbolicheskikh sistem uravnenii* (M.: Fizmatlit, 2001)
8. R. Hockney, J. Eastwood, *Computer Simulation Using Particles* (New York: Adam Hilger, IOP Publication Ltd, 1989)
9. C. Birdsall, D. Fuss, J. Comput. Phys. **135**, 141 (1997)

The Convergence of the telematic, computing and information services as a basis for using artificial intelligence to manage complex techno-organizational systems

Alexander Raikov[1], *Arkady* Trachuk[2], *Yulia* Romanova[2], *Valery* Loginova[3,*], and *Vitalia* Bortalevich[4]

[1]RAS, Institute for Control Sciences, 117218 Moscow, Russia
[2]Financial University under the Government of the Russian Federation, Department of Management, 125993 Moscow, Russia
[3]RAS, Institute of Market Problems, 117418 Moscow, Russia
[4]National Institute of Energy Security, 119992 Moscow, Russia

Abstract. The authors analyses the problems of using artificial intelligence to manage complex techno-organizational systems on the basis of the convergence of the telematic, computing and information services in order to manage complex techno-organizational systems in the aerospace industry. This means getting the space objects a higher level of management based on the self-organizing integration principle. Using the artificial intelligence elements allows us to get more optimal and limit values parameters of the ordinal and critical situations in real time. Thus, it helps us to come closer to the limit values parameters of the managed objects due to rising managing and observant possibilities.

1 Introduction

The main advantage of using the artificial intelligence to manage complex techno-organizational systems on the basis of the convergence of the telematic, computing and information services is having more opportunities to collect, process, keep and distribute information [1]. In other words, this means the ability to adapt to the ordinal and critical situations dynamics [2, 3].

In this way such using the artificial intelligence improves the management system by using more simple and faster planning, setting and controlling space objects which form quasi-unified system [4-6].

The realization of the self-organizing integration principle may be applicable to the various kinds of technical problems. For example, it can be used in constructing the space power plants from the little standard modules launched through electromagnetic accelerators.

* Corresponding author: loginovalerochka@mail.ru

The description of each type of identifiable electronic control transactions performed on systems in a network (cloud, swarm) of standard modules as elements of a space power plant contains a specific set of attributes for it.

Descriptions of identifiable electronic control transactions (through dedicated clusters of managed space objects) associated with the identification and analysis of atypical situations of the same type can be grouped into information blocks that represent virtual information blocks for the described schemes for implementing self-organizing integration processes.

2 Problem statement

Instrumental support of the processes of analyzing the information resources of objects participating in the processes of self-organizing integration, available for monitoring to structures that implement the functions of automated monitoring and control over the effective and safe operation of complex techno-organizational systems in the aerospace industry in the process of self-deployment in the network orbit (clouds, swarms) of standard modules as elements of space power plants is as follows.

Integration of information systems of various groups of controlled space objects is provided by standardizing the descriptions of the identifiable electronic control transactions made with respect to systems in the network (cloud, swarm) of standard modules as elements of a space power station when space power stations are built in orbit from small standard modules transceivers and microcontrollers) triggered by electromagnetic accelerators that support the modes of self-organization the integration of a swarm of controlled space objects and the support in each of them of a standard protocol.

Functional relations determine the nature of content formation for automating the procedures for constructing both concepts and denotata, where the basic characteristics of monitoring are established to maintain the regimes of self-organizing integration of a swarm of controlled space objects (including the identification of cause-effect relationships between the dynamics of changes in the modes of operation of the system) renewable and dynamically adapted to the individualized profile modes of self-organizing integration of a swarm of controlled space objects ect.

3 Solution

On this basis, it is possible to identify the technological activity of controlled space objects creating increased risks, where the behavior of the participants in technological interaction makes it possible to distinguish the organizational strategy of controlled space objects creating increased risks and their characteristics (chronotype, participants, role structure, etc.).

For example, the analysis of technological schemes with the identification of relevant emergency space objects and an assessment of their significance on the basis of statistics of accidents and incidents with different profiles of objective activity in the network (cloud, ro) of standard modules as elements of a space power plant.

The structural relationships determine the identifiable links of operational-regime situations dynamically localized as temporary, resource or any other matrix that formalizes the information portrait of the process of maintaining the regimes of self-organizing integration of a swarm of controlled space objects using structured content blocks of distributed databases consisting of different data, versions or structured options, typical or atypical.

Affiliation ties determine the affiliation of information packets distributed over computational nodes as elements of a dynamic computing cluster, including structured information materials for building stable operational and communication links of controlled space objects with a common organizational information platform and a coordinated information exchange management system for managing a complex of modules ("sides"), .

The core of the coordinated information exchange management system should be a converged information and computing platform, which simultaneously forms the driver for the network interaction of other computing nodes as elements of a dynamic computing cluster.

Such a system should combine information, telematic and computing services for the development of information systems for managing a swarm of controlled space objects as elements of a quasi-unified system.

While conducting monitoring on information and technological chains of information management, monitoring, storage and exchange creates a kind of dynamically changing multilayered network of semantics of analyzed links between identified situations in the processes of self-organizing integration.

Each layer of such a network corresponds to a certain class of links mediating the provision of the concentration of all information on the passage of each process of self-organizing integration and all together, on the ongoing transactions and participants in technological interaction for self-deployment of standard modules in the orbit of the network (cloud, swarm) as elements of space power plants.

The possibilities of network-centric integration of data with respect to the content of distributed databases, consisting of data received from space and ground objects, provide communication between the monitoring and control operators during the implementation of the whole range of different technological processes for self-deployment in the orbit of the network (clouds, swarms) modules as elements of space power stations.

The detection of the described system-dynamic communication serves as a message to the monitoring and control operator about a new connection within the analyzed distributed databases and test results with the formation of a temporary, resource or operational-regime matrix of the process of self-organizing integration that are carriers of information determined by the class and description of this connection .

The possibility of forming a temporary, resource or operational-regime matrix of the studied process of self-organizing integration is created, which can be based both on a separate event (the result of a technological operation) and on aggregation and comparison of dissimilar technological events and network flows of such operations.

In this complex, a unique opportunity is created for a principled increase in the effectiveness of the organizational mechanisms for reducing the risks of accidents and improving the sustainability of the control processes for the swarm of controlled space objects in direct interaction between the board and the aircraft, or when transmitting on board each module information from the ground-based surveillance system.

The complex makes it possible to detect the interrelationships between controlled space objects, actors, events and various implicit correlations between them with respect to the detection, based on analysis of monitoring results, of the explicit and latent characteristics of the process of providing self-organizing integration (formation of the core of an intelligent generator and energy translator, integrable from separate quasi-autonomous elements in a network (cloud, swarm) of standard modules) and its correspondences normative requirements.

As a result, there is a distributed-network formation of a set of measures to prevent emergencies, situations with non-fulfillment or inadequate execution of regulations while maintaining the regimes of self-organizing integration of a swarm of controlled space objects with a constant step-by-step comparison with the required normative model of using complex techno-organizational systems in the aerospace industry for self-deployment at Orbit network (cloud, swarm) of standard modules as elements of space power plants.

To ensure the safety and effective use of complex techno-organizational systems in the aerospace industry in the process of self-organizing integration, a method of intellectual analysis of the fractal organization of key element relationships is proposed, using a multidimensional interpretation of the linkage systematics and operational-regime control transactions.

The new technology provides a dynamic study of the semantics of the analysis of explicit and latent relationships in databases containing the necessary information about operational-regime situations.

4 Conclusion

Automated monitoring and control is performed to determine the basic characteristics of the system-dynamic analysis of the electronic content of control transactions while maintaining the regimes of self-organizing integration of a swarm of controlled space objects, testing - the time, resource or operational-mode matrix of any studied self-organizing integration process energy from various points of geostationary orbits of space solar power stations on the ground.

The quasi-unified system of space object management has some principal differences from the traditional one and that makes it substantially new one.

In order to improve the functioning of the system it is necessary to add the next elements to the using artificial intelligence process:

1) The monitoring subsystem of the measurable parameters of the various kinds of physical fields, environments and objects which provide the forming of the current space objects modules situation.
2) The diagnostics of anomalies subsystem that will identify the anomaly`s reasons and determine the current situation on controlled objects.
3) The analysis of the current indicators subsystem which will evaluate the functionality of stabilization process of the space objects (including adding and reducing of the elements)
4) The subsystem of evaluation of the current plan of the telematic, computing and information services management which analyzes the working system and warns if there are problems.
5) The intellectual agent which is planning and correcting the work of the telematic, computing and information services.

Application part should include the following modules:

- the real-time monitoring of elements and processes;
- the modeling, recognizing and analyzing system of the global and current situation and supporting decision-making and in the ordinary conditions;
- the system for predicting problems in real time;
- the intelligent real-time agent for correcting the management and optimizing solutions;
- the interface for making recommendations to the modules in a critical situation.

The theses were prepared with the financial support of the Russian Humanitarian Scientific Foundation (project No. 16-02-00463 a "Formation of organizational mechanisms for operating oil and gas resources on the basis of multi-agent modeling to protect Russia's economic interests from manipulating oil prices on world markets").

References:

1. A. S. Bugaev, E. L. Loginov, A. N. Raikov, V. N. Saraev, S&T. Inf. Pr. **36 (1)**, 68 (2009)
2. E. L. Loginov, A. N. Raikov, Therm. Eng. **62(4),** 233 (2015)

3. A. P. Grigorev, A. I. Soldatov, P. V. Sorokin, *MTT*, 97 (2000)
4. D. N. Demyanovich, O. S. Vadutov, A. I. Soldatov, MEACS, (2014)
5. A. S. Asochakov, Y. V. Shulgina, A.I . Soldatov, E. M. Shulgin, D. N. Ogorodnikov, MEACS, (2015)
6. S. V. Chubov, A. I. Soldatov, IOP Conf. Ser. **177(1)**, (2016)

High speed video recording system on a chip for detonation jet engine testing

Alexander N. Samsonov*, and *Khristina* V. Samoilova

[1]Lavrentiev Institute of Hydrodynamics SB RAS, 630090 Novosibirsk, Russia

Abstract. This article describes system on a chip development for high speed video recording purposes. Current research was started due to difficulties in selection of FPGAs and CPUs which include wide bandwidth, high speed and high number of multipliers for real time signal analysis implementation. Current trend of high density silicon device integration will result soon in a hybrid sensor-controller-memory circuit packed in a single chip. This research was the first step in a series of experiments in manufacturing of hybrid devices. The current task is high level syntheses of high speed logic and CPU core in an FPGA. The work resulted in FPGA-based prototype implementation and examination.

1 Introduction

High speed video recording device bases on a sensor equipped with a large number of parallel data buses, each of which calls "channel". Data is transmitted over up to 16 channels in parallel clocked by the same clock wire. The channel width is equal to ADC output width which is usually 10 or 12 bits. However, common data width largely exceed 160 bits and it does not allow to use general purpose CPUs.

The first prototype was constructed using FPGA, because it is suitable for debugging and verifying of architecture and firmware. During architecture development the number of FPGA in device was increased to four. Each of FPGA conducts up to four 12 bit data channels and it controls external DDR3-memory. Incoming data is transmitted through internal FIFOs to realize cross domain clock synchronizations. There are a lot of clock domains because data width varies in different interfaces. For example memory width is 24-bit, external Gigabit interface width is 8-bit and input data width is up to 72 bit for single FPGA.

2 Arithmetic

High-speed video recording process is a memory intensive routine, which demands real time data analysis. Also sensor calibration procedure is performed during data analysis. Calibration coefficients are stored in internal memory. Normalized and linearized data is fed through outline recognition module and image recognition is performed.

* Corresponding author: info@ck-llc.com

Implemented core includes following arithmetic and logical functions tested in FPGA and applied for image recognition [1]:
- averaging;
- differentiating;
- segmentation;
- hysteresis based on previous frame and threshold level;
- pixel data confidence.

Some of these algorithms demand too many multipliers. Some of them are memory intensive. This property was important during FPGA family selection. So multipliers in Xilinx Virtex 5 was organized in columns (in sequence of) 64 pcs, but delay between different columns if you connect it serially was too big [2]. Altera FPGA does not have this problem: all off multipliers in Stratix V can be connected serially. But, nevertheless, number of multipliers depends on FPGA family and price. ASIC implementation will help create appropriate sequence of multipliers connected in according to minimal delays.

3 FPGA architecture

3.1 Clock-domain synchronization

Different modules and interfaces has different clock frequencies. So clock domains synchronization routine implementation was necessary. First In First Out (FIFO) blocks were used for this purpose. In Figure 1 structure of data storage, transformation and transmission modules for FPGA based set up is shown. There are two independent clock sources, one of which follows sensor-emitted data, the second – is high speed external memory driven base clock.

Every module is driven by finite state machine (FSM), which checks for data ready state, presence of data request and prevents output buffer underflow and overflow. Block diagram is shown in Figure 2.

3.2 Data queue delay

Different data transmission modules employed in design have different number of synchronization stages in queues. As a result, input FIFO data receiving from high speed external memory is based on a four-stage synchronization queue, but output FIFO data transmitting to Ethernet interface has a single stage queue. It influenced output data delay in a way it was eight clock cycles in first FIFO and two in the second. The problem in FPGA synthesis appeared because as soon as logic utilization exceeds 80% and some calculations are memory and logic intensive router faced at problem of too long interconnection wires. Because of it one of the signals will rise too late because of too long data path. This issue has to be taken into account during selection of an algorithm. Writing of additional synthesizer constraints is necessary if logic utilization is too high [3]. This constraints are described in Synopsys design constraints file, so that it can be converted easily into ASIC constraints file. It will help to optimize netlist structure and quality of routed topology [4].

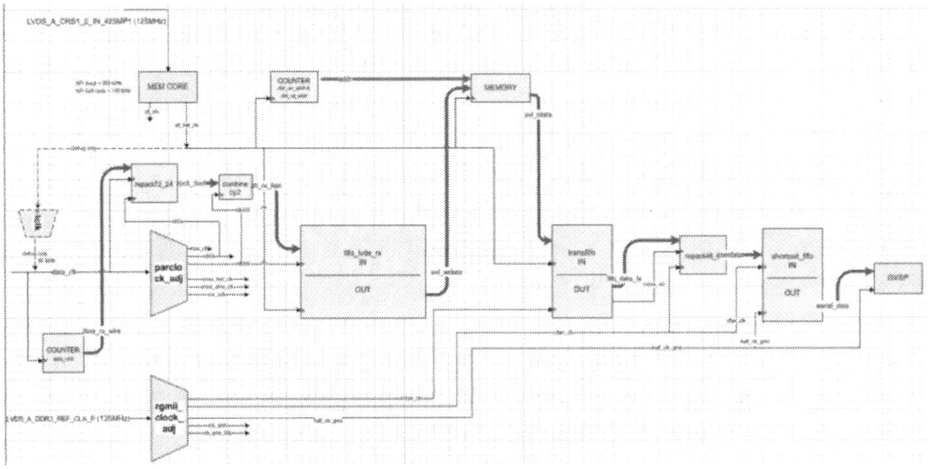

Fig.1. Structure of data transmission modules.

3.3 Optimization, bit shift repairing, data recovery

Incoming data is synchronized to data valid signal which was fed from sensor. However so many transformations caused in unclear placement of start line and start frame pixels. Start of line and start of frame markers were added to fix this issue. It helped to restore image borders due to marker placement.

Input data was synchronized according to the clock phase. Dynamic phase alignment routine was implemented for this purpose. Algorithm was based on finding a template sequence in data stream and shifting the data clock in according to the template reference clock carrier's frequency and phase. It chooses clock phase restoring template sequence correctly and use it to shift data reference clock.

3.4 Paralleling

All FPGAs were based on partly the same firmware (depicted in Figure 1). Some differences were in pin assignments of chip to chip data exchange modules, because ones are master and the others are slave devices. Master devices had external interface module and transmitted data over Ethernet connection.

4 ASIC implementation

4.1 ASIC advantages

To exclude cyclic transmission of incoming data through multiplier queue and to reduce FIFO number and bus width transformation architecture was specially adjusted for ASIC implementation [5, 6].

After adjustment has been finished RTL code was compiled into logic primitives (netlist) and routed in silicon CMOS-structure.

4.2 Architecture

ASIC can not be reconfigured flexible against FPGA so additional mechanisms of function adjustments should be implemented. In according to this purpose control and status registers were employed. Moreover to set values into registers and control data to memory and Ethernet transactions RISC-V CPU-core was implemented. CPU based on Syntacore SCR1 realization written in System Verilog. Data was transmitted internally over 32-bit AXI4-bus. It provided data control logic and it manages access to data bus, check whether it is busy or not. Ethernet medium access controller (MAC) was implemented and routed over AXI4 bus to CPU core. GigEVision interface core was implemented in Verilog and connected to logic unit and external interface so that it works like intermediate chained routine. See Figure 3 for precise architecture description.

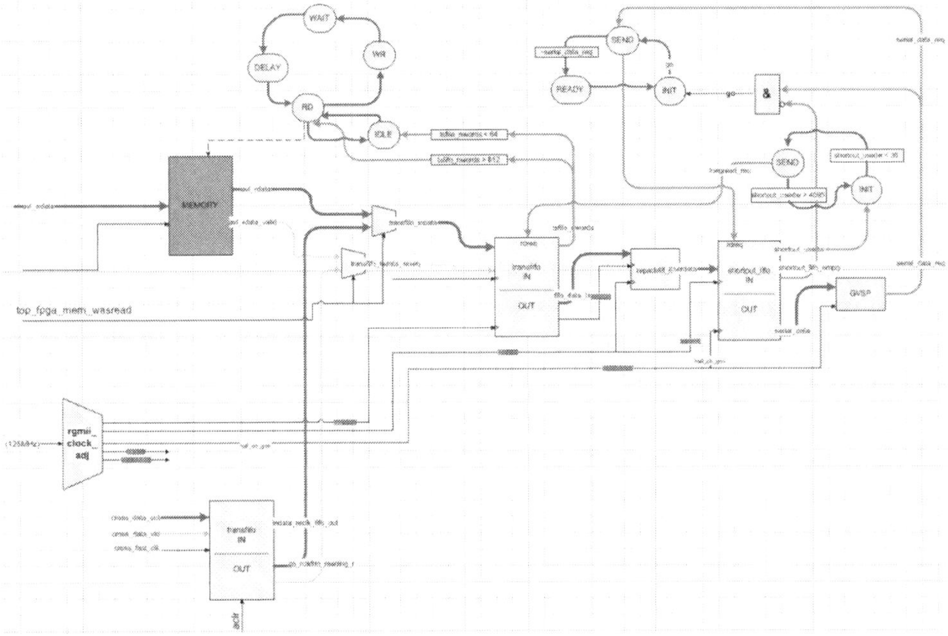

Fig. 2. Finite state machine placement and its role.

In result ASIC implementation can exclude FIFOs and difficult cyclic calculation due to increased buses width and multiplier queues. Also operating frequency will be increased and the demands on technological process precision will be considerably reduced.

4.3 ASIC netlist implementation

During FPGA debugging synthesis constraints were adjusted initially and it was the starting point to adjust ASIC synthesis constraints, which described timing and mapping restrictions. Also drive strengths and slew-rate settings to top-level ports of the design were determined. Timing-driven I/O register mapping was performed. Number of nodes driven by top level port instance was under control – fan-outs were restricted. This values were written in Synopsys Design Constraints (SDC) format. During RTL synthesis behavioral RTL code was transformed into structural RTL code according to technological cell libraries. In result behavioral Verilog file was converted into logic level Verilog implementation[8-11].

Additional netlist synthesis tools (RTL schematic, technology-mapped schematic, and critical path schematic) helped to make further constraints to optimize the design. In result technology-mapped netlist was generated.

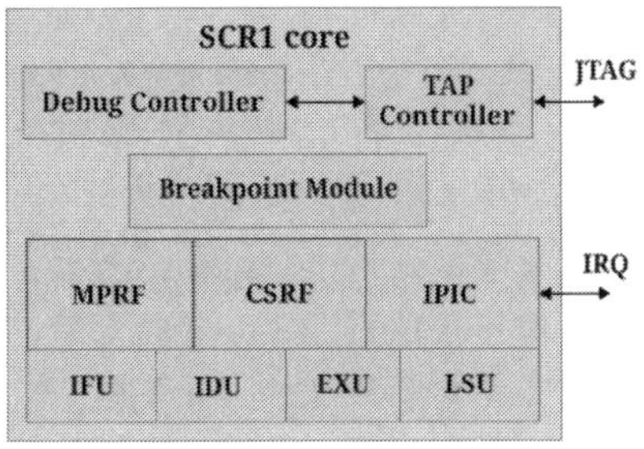

Fig. 3. CPU modules structure.

5 Conclusion

Signal analysis algorithms developed for high speed camera were tested in FPGA and are ready to be implemented in silicon. It helped to perform image transformation in real time and decreased overall power consumption and data transmission on request delay during starting of device and after exposition is finished. Some data path bottlenecks were removed in according to parallel architecture optimization technique. In result power consumption was reduced and logic stability was improved[7, 12-13]. This factors helped to use device in different areas such as propulsion engine testing [14], aircrafts, observation balloons and space satellites.

References

1. I. A. Kozulin, O. V. Nikanorov, S. N. Koreshev, J. Optical Tech **75(9)**, 558 (2008)

2. A. N. Samsonov, A Pat. Rec. and Im. Analysis: A. M. T. and A. **25(2)**, 255 (2015)

3. A. N. Samsonov, F. A. Bykovskiy, *Int.Conf. on M/Nanotech. and El. D.,* 160 (2011)

4. T. V. Perevalov, D. R. Islamov, A. A. Saraev, Tech. Phys. Letters **42(6)**, 601 (2016)

5. T. V. Perevalov, V. S. Aliev, V. A. Gritsenko, A. A. Saraev, V. V. Kaichev, Microelectronic Engineering **109**, 21 (2013)

6. D. R. Islamov , V. A. Gritsenko, T. V. Perevalov, O. M. Orlov, G. Y. Krasnikov, APL **109(5)**, 052901 (2016)

7. D. I.Rogilo, N. E. Rybin, L. I. Fedina, A. V. Latyshev, Optoelectr. Instr. D. Process. **52(5)**, 501 (2016)

8. A. G. Pogosov, M. V. Budantsev, E. Yu. Zhdanov, D. A. Pokhabov, A. K. Bakarov, A. I. Toropov, APL **100**, 181902 (2012)

9. K. V. Feklistov, A. G. Cherkov, V. P. Popov, Solid State Com. **242**, 41 (2016)

10. A. A. Lyamkina, D. V. Dmitriev, A. I. Toropov, S. P. Moshchenko, APL **110(1)**, 011103 (2017)

11. A. Regler, K. Schraml, M. Spiegl, K. Müller, J.J. Finley, M. Kaniber, J. Vuckovic, A.A. Lyamkina, J. Nanophotonics **10(3)**, 033509 (2016)

12. D. I. Rogilo, L. I. Fedina, S. S. Kosolobov, A. V. Latyshev, B. S. Ranguelov, Phys. Review L. **111(3)**, 036105 (2013)

13. D. I. Rogilo, L. I. Fedina, S. S. Kosolobov, A. V. Latyshev, B. S. Ranguelov J. Crystal Growth **457**, 188 (2017)

14. Fedor A. Bykovskii, Sergey A. Zhdan, Evgenii F. Vedernikov, J. Prop. and Power **22(6)**, 1204 (2006)

Analysis of methods for measuring the liquid level in the annular space of an oil well

Juriy Shinyakov, *Maxim* Sukhorukov, *Daria* Torgaeva, *Andrey* Soldatov*, *Natalia* Shalyapina, and *Dmitriy* Li

Tomsk State University of Control Systems and Radioelectronics, Research Institute of Space Technology, 634050 Tomsk, Russia

Abstract. One of the most important diagnostic parameters of an oil well operation is the dynamic liquid level in the annular space. In paper the main methods for determination of the dynamic liquid level that does not require suspension of the well work and its depressurization are considered. Also the comparative analysis of these methods is given.

1 Introduction

Currently, oil production is carried out using sucker rod pumping units (SRPU) and electrical submersible pumping units (ESPU). Most of Russian's production well is equipped with SRP, despite the great amount of disadvantages (bulkiness of the above-ground part, incomplete wellhead sealing, limited capacity, etc.). It is also worth noting that SRP is the most suitable unit for oil production in difficult conditions (deep bedding, sand or paraffin admixture, high gas factor) which is highly topical for using in depleted wells [1].

The dynamic liquid level in the annular space of the oil well has a huge influence on the SRP operation. If the operation mode of the pump does not correspond to the rate of flow of liquid from the reservoir a decrease of dynamic level may be observed, which will lead to pressure release at the pump intake. In this case, oil degassing will lead to incomplete filling of the pump and a decrease of pump slippage. A further dynamic level decrease will lead to pump starvation. If the emergency shutdown does not actuate the operation of the pump in dry friction mode will lead to wear and damage of the parts [2, 3].

It is recommended to increase the depth of bore-hole pump to a dynamic level in the case of high gas con-tent in the oil-water mixture. This will lead to a pressure boost on the pump intake and a large part of the gas in the oil will be in a dissolved state. This fact will positively affect the production of the SRP. However, due to an increase in the weight of the rod string and acting floating force the mechanical and hydrodynamic resistance may increase. As a result, delivery rate and pump efficiency will be more reduced [4, 5].

The fluctuation of the liquid level in the annular space also has a negative effect on the pump operation because there are load variations at all nodes of the unit, rapid wear and damage of the SRP components.

There are the following methods for measuring the dynamic liquid level:

* Corresponding author: soldatov.88@bk.ru

- Echometry and wavemetering;
- Thermomanometric system;
- Dynamometry and wattmetry.

2 Echometry and wavemetering

Echometry and wavemetering are one of the most popular methods for determining the dynamic liquid level of fluid in the annular space of an oil well. They are based on the calculation of the product of the acoustic velocity in a gaseous medium for the transit time of the signal from the emission source (wellhead) to the liquid level and back. The echometry is used for measurements in low-pressure wells, and wavemetering for wells with high pressure.

However, using this method you can get an equivocal result. Firstly, it is difficult to recognize the actual position of the liquid level on the echogram because there are noises at the receiver input. The vibrations of the tubing string, the reflection of the signal from heterogeneities in the annular space, acoustic noise of the pump also affect the acoustic signal. For this reason, the useful signal can be indistinguishable from interference. Secondly, the acoustic wave is reflected not only from the surface of the liquid, but also from any border line with a significant change in density. It is proved that at a gas-liquid mixture density of 200 kg / m3 it is sufficient to reliably fix the response. For example, a foam column is often formed on the surface of a liquid and is a gas-liquid mixture whose density increases in depth [4]. Thirdly, the acoustic velocity in the annular well space is not constant, its value depends on the temperature, pressure, density and composition of the gas and can vary from well to well with same annular pressure even within one deposits.

Despite the disadvantages, the echometry is the main one in the oil industry because it is fairly simple to implement and inexpensive. A lot of works are devoted to the study of various techniques and tools that increase the accuracy of liquid level measurements by echometry [5-12]. For example, in algorithms for estimating the re-cording time of reflected signals are considered, structural schemes of pulse formers for sounding the annular space are proposed, which make it possible to obtain a more powerful signal that can be distinguished against background noise. Works [13, 14] describe methods for determining the speed of sound in a well: design, pipe, and benchmarking.

3 Thermomanometric systems

This method involves placing pressure and temperature sensors along the wellbore at a distance of 100 m from each other. Serial interrogation of sensors is performed to determine the dynamic liquid level. After recording the coordinates of the point "depth, pressure" in an array corresponding to the region of the well filled with gas, the correlation coefficient of the points of the entire array is calculated by least square method. The operation is repeated until the value of the coefficient does not change by more than 10% with respect to its previous value. This point indicates that the current sensor is immersed in the liquid. From now on, the "coordinates" are written to another array corresponding to the area of the well filled with liquid. After conducting a survey on the data recorded in the arrays, 2 graphs of the pressure dependence on the depth of the sensors are constructed. The point of intersection corresponds to the level of the liquid (Fig. 1).

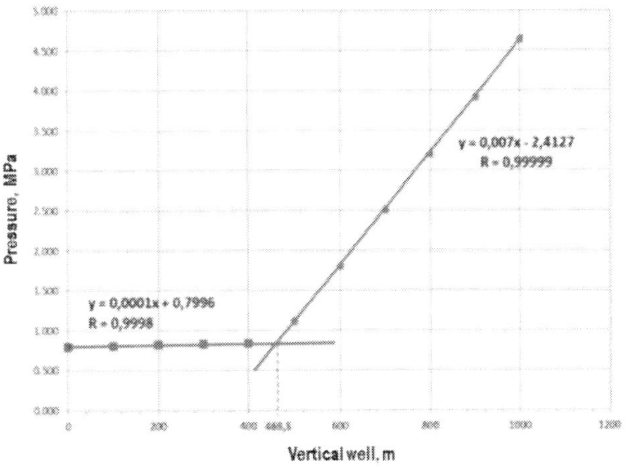

Fig.1. Determination of the liquid level in the well.

The authors of [13, 14] proposed to place pressure sensors along the armored vertical cable. This method is the perfect approach for ESPU (because the pump is located at the bottom of the unit, an insulated power cable leads to it). The thermomanometric system makes it possible to estimate the state of gas and liquid in the annular space in real time, however, the main disadvantage is the high cost.

A similar method for determining the dynamic level is described in [15]. Temperature sensors must be located along the entire length of the well at a distance of at least 20-40 m from each other. After polling all the temperature sensors, a thermogram is formed, along which the temperature gradient at each point is calculated. The point with the maximum value of the gradient will be the border line (Fig. 2).

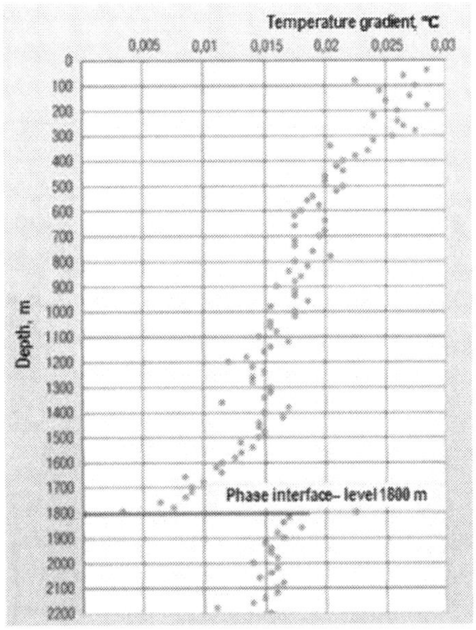

Fig. 2. Determination of the liquid level in the well according to the temperature gradient.

4 Dynamometry and wattmetry

There are methods to determine the dynamic liquid level by dynamometer cards (the dependence between load on the polished rod and its stroke) and wattmeter cards (the dependence between power consumed by the motor and time).

Based on dynamometer card, it is possible to determine not only the load on the polished rod, but also the production rate of the well, and the condition of the pumping unit as a whole. The dynamometer card is the result of the interaction of a huge number of factors, so its interpretation is a very difficult task. Based on the deviation of a real dynamometer card from an ideal mathematical model, any deviations from the norm are fixed.

The article proposes a technique for calculating the change in the dynamic liquid level by an ideal dynamometer card. Initial data are the parameters of the pump itself, as well as the depth of its immersion and the value of the pressure in the pipe. This estimation method is indirect and requires large computing powers. However, the dynamic level change is usually monitored to prevent pump failure. The ingress of gas into the pump cylinder causes significant changes in the shape of the dynamometer card. In particular, when the pressure on the pump intake valve decreases, a dynamometer card of the form shown in Fig. 3a goes to the form Fig. 3b. This is caused by large gas content in the oil or a decrease in the dynamic level before the pump is received.

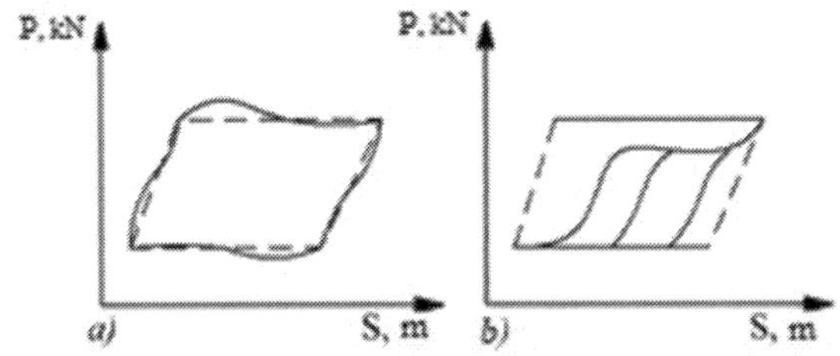

Fig. 3. Practical dynamometer cards of the SRP operation: a) normal operation with a small number of oscillations, b) gas effect.

Dynamometer cards and wattmeter cards are connected by complex non-linear dependencies. Determining the dynamic liquid level and other parameters of the pump by using wattmeter cards is an urgent task. These data allow to control the operation of the pumping unit according to the indications of a single sensor (almost all SRP are equipped with power sensors). However, a significant disadvantage of this method is the complex process of interpretation the wattmeter cards.

5 Conclusion

As a result, he most promising method from the considered is the method of determining the dynamic liquid level on wattmeter cards because it does not require the installation of additional equipment and allows to determine the state of the entire pumping unit by one single parameter.

The work was supported by the Ministry of Education and Science of the Russian Federation, Agreement No. 14.574.21.0157 (unique identifier RFMEFI57417X0157), within the framework of the FTP project "Research and Development in Priority Directions for the Development of the Russian Science and Technology Complex for 2014-2020".

References

1. F. Terraneo, L. Rinaldi, M. Maggio, A. Papadopoulos, A. Leva, *P.R.T. Sys. Symp.* 7010370, 11 (2015)
2. S. K. Phang, J. J. Ong, R. T. C. Yeo, B. M. Chen, T. H. Lee, *SIBIRCON*, 722 (2010)
3. V. D. Yurkevich, *SIBIRCON*, 124 (2015)
4. A. I. Soldatov, A. A. Soldatov, S. I. Bortalevich, O. A. Kozhemyak, P. V. Sorokin, E. L. Loginov, Y. A. Shinyakov, M. P. Sukhorukov, *SIBCON*, (2017)
5. Y. V. Shulgina, A. L. Starostin, M. A. Kostina, T. S. Mylnikova, A. I. Soldatov, MEACS, (2015)
6. A. I. Soldatov, O. A. Kozhemyak, A. A. Soldatov, Yu. V.Shulgina, IOP Conf. S: Mat. S&E, (2015)
7. Y. V. Shulgina, A. A. Soldatov, E. M. Shulgin, A. V. Kudryashova, *SIBCON*, (2015)
8. Yu. V. Shulgina, A. I. Soldatov, Ya. V. Rozanova, A. A. Soldatov, E. M. Shulgin, IOP Conf. S: Mat. S&E, (2015)
9. Y. V. Shulgina, A. I. Soldatov, E. M. Shulgin, Y. V. Rozanova, M. Kroning, MEACS, (2014)
10. A. I. Soldatov, A. I. Seleznev, A. A. Soldatov, P. V. Sorokin, V. S. Makarov, Rus. J. NDT **48(5)**, 268 (2012)
11. A. I. Soldatov, A. I. Seleznev, A. A. Soldatov, I. I. Fiks, X. M. Kroening, Rus. J. NDT, **48(4)** 255 (2012)
12. A. I. Soldatov, J. V. Chiglintseva, *SIBCON*, (2009)
13. H. Bodur, A. F. Bakan, M. A. Baysal, E. E. **85 (1)**, 45 (2003)
14. M., Borage, S. Tiwari, S. Kotaiah, IEEE T. on I. Electr. **52 (6)**, 1547 (2005)
15. Y. A. Shinyakov, V. D. Semenov, V. A. Kabirov, D. S. Torgaeva, M. P. SuKhorukov and R. S. Cevastyanov, SIBIRCON, 346 (2017)

Digital control systems for power supplies of space vehicles

Andrey Soldatov *, Juriy Shinyakov, and *Maxim* Sukhorukov

TUSUR, Space Technology Institute, 634050 Tomsk, Russia

Abstract. The principle of building fault-tolerant digital control systems for power supplies of space vehicles is considered in the article. The analysis of the circuits by sampling the controlled component of the output voltage at discrete instants of time is carried out. The possibility of using this approach in multi-channel digital control systems is shown. The results of theoretical studies are confirmed by the example of a four-channel control unit.

1 Introduction

Widely used analog control systems for switching voltage regulators (SVR) have a number of significant disadvantages: insufficient flexibility in control, impossibility of implementing complex control algorithms, low noise immunity and limited accuracy due to leakage currents, zero-point voltages, temperature of parameter instability, etc.

The use of digital control systems (DCS) improves the parameters of the SVRs, which will provide the following opportunities:

1) selection of the optimal operating modes of the power elements (switching frequency of power switches, change of induction in magnetic elements, current ripple in inductive elements and voltage across capacitors, etc.) [1];
2) the introduction of mathematically complex transfer functions (control laws), their modification without redesigning the equipment;
3) maintenance and issuance of signal information on the load, primary power supply, dangerous operating modes, etc.;
4) realization of intellectual functions of adaptation, self-diagnostics and self-adjustment;
5) forecasting catastrophic regimes;
6) the organization of any interfaces for modern consumers and many other functions [2-4].

Digitization of control algorithm makes it possible to implement the most promising predictive algorithms (predictive control) of frequency control by the pulse width modulation (PWM), which allows reducing the delays in the control loop and the delays of information transfer.

2 Problem statement

* Corresponding author : soldatov.88@bk.ru

When developing PWM switching voltage regulator in power supply systems (PSS) of space vehicles (spacecraft), the task of researching dynamic characteristics and forming optimal control laws in the "small" signal mode is put on the foreground. Using the approach proposed in [5], it is possible to study dynamics from continuous models, and quantization errors can be estimated from the stationary periodic regime. However, at the synthesis of systems using such continuous models the possibilities of pulse regulation are not used. Other approaches involve taking into account the discrete nature of processes in PWM systems by means of data reduction to the models with pulse-amplitude modulation (PAM) [6], or to the models with a variable structure in the state space [7].

3 Problem-solving methods

Of all the known approaches to the synthesis of control laws that ensure high speed with stabilization of the output voltage, a special method is of interest:

- representation of the power circuits of the switching voltage regulators with PWM in the mode of small deviations as an adequate amplitude-impulse model with the controlled components of the processes;
- synthesis of the control law using polynomial equations for the systems with PAM;
- implementation of the synthesized control law in switching voltage regulator taking into account specifics of PWM.

This method allowed its authors to synthesize a control law that provides the minimum finite duration of transient processes in the switching voltage regulator based on DC-DC buck converter. Such a law is called optimal by the authors for speed, and a voltage regulator with such a control law is a high-speed voltage regulator. In the proposed approach to the synthesis of the control law was further developed for both boost and buck-boost DC-DC converters.

In accordance with the discrete transfer function of the synthesized correction device and taking into account the specifics of PWM, two ways to implement the PWM control device are proposed:

- on instantaneous values of state variables;
- on the values of state variables, selected at discrete instants of time.

$$W_K(p) = d + d\left(1 - e^{-pT}\right) \quad (1)$$

where $d = LC/T$, L and C are the inductance and capacitance of the output filter of switching voltage regulator.

The implementation of the *PWM* control law, in which only discrete values of the regulated component of the voltage across the capacitor of the output filter are used, is shown in Fig. 1.

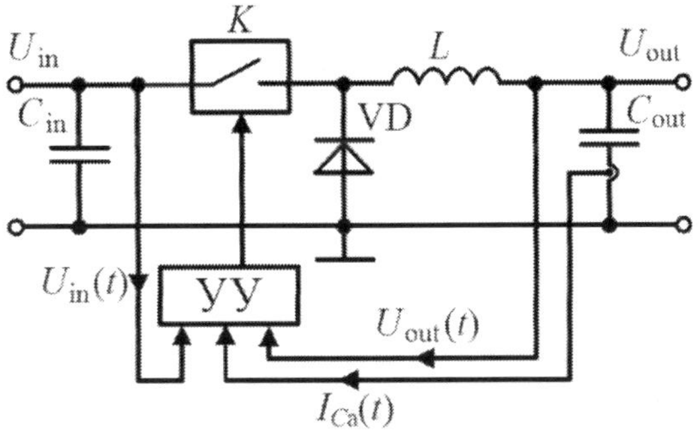

Fig 1. Switching voltage regulator with discrete control

In this scheme, the increment of the input signal of the pulse-width modulator is determined from the expression:

$$\Delta U_y(mT) = -\frac{d_0}{U_{in} K_m}\left[2\Delta U_{Cp}(mT) - \Delta U_{Cp}(m-1)T\right], \tag{2}$$

where U_{in} is the voltage at the input of the scheme; $K_m = \Delta t_{c.p} / \Delta U_y(mT)$, $\Delta t_{c.p}$ - increment of the pulse duration of the control pulse of the power switch of the regulator with respect to the time mT of the controlled switching; $\Delta U_y(mT)$ is the increment of the *PWM* input signal; Um is the amplitude of the sawtooth PWM voltage, *T* is the conversion period.

Given the condition $\Delta t_{c.p} < T$, it can be assumed that the controlled switching of the power switch of the voltage regulator occurs with a period of *T*.

The current values of the dynamic component of the input signal of the pulse-width modulator are given by:

$$U_y(mT) = U_y((m-1)T) + \Delta U_y(mT), \tag{3}$$

It is impossible to determine the regulated component of the voltage U_{Cp} across the capacitor of the output filter of the regulator by performing computational operations with discrete or continuous values of the output voltage U_{out} of the scheme. This is explained by the fact that the equivalent circuit for the capacitor of the output filter can be represented as a series connection of capacitance C_f and internal resistance R_C of a capacitor. Therefore, the output voltage of regulator which is the voltage across the capacitor of the output filter of scheme consists of the voltage across the capacitance U_C and the voltage across the internal resistance U_{Rc}.

Since the internal resistance R_C of the capacitor can significantly change under the influence of temperature and time factors, then the voltage U_{Rc} will also change, and therefore, to determine the voltage across the capacitance U_C for example, by subtracting the undefined voltage U_{Rc} from the output voltage U_{out} is not possible.

It is possible to calculate the increment of the regulated component of the voltage ΔU_{Cp} across the capacitance over the period *T* by integrating the increment of the regulated component of the capacitor current ΔI_{Cp} in the time interval equal to the period *T*:

$$\Delta U_{Cp}(mT) = \frac{1}{C}\int_{(m-1)T}^{mT} \Delta I_{Cp}((m-1)T)dt, \tag{4}$$

Since the increment of the regulated component of the capacitor current in the interval between the adjustable switching times *mT* of the power switch of the voltage regulator remains unchanged (Fig. 1), then to determine it the first difference in the capacitor current should be calculated:

$$\Delta I_{Cp}(mT) = I_{Cp}(mT) - I_{Cp}((m-1)T)$$

or

$$\Delta I_{Cp}(mT) = I_{Cp}(mT + \tau) - I_{Cp}((m-1)T + \tau), \tag{5}$$

where $\tau < T$ is, in general case, an arbitrarily chosen fixed time interval.

In order to have the time for carrying out the computational procedures necessary to determine the input signal of the pulse-width modulator at the instant of time mT, it is advisable to select τ such a way that the moments of time $(m-1)T + \tau$ are as far removed from the moments mT of the adjustable switching of the power switch of voltage regulator. For DC-DC buck converter and for modulation of the fall time of the pulse, the time instants $(m-1)T + \tau$ should be selected immediately after the power switch is turned on.

As a rule, the static duty cycle $K_{St} > 0.25$ in the buck DC-DC converters, therefore, the sampling time of the discrete values of the input signals and the computation is still close to a quarter of the conversion period. Using expression (1.5) allows us to define $\Delta I_{Cp}(m-1)T + \tau$ at time $(m-1)T + \tau$.

Since the increments of the regulated component of the voltage in the interval between the adjustable switching times of the power switch of the voltage regulator are constant, then (1.4) can be written as:

$$\Delta U_{Cp}(mT) = \frac{1}{C} \Delta I_{Cp}((m-1)T), \tag{6}$$

Thus, replacing the integration procedure in accordance with (1.4) by determining the area of the rectangle according to (1.6) allows us to determine the increment of the regulated component of the voltage across the capacitance $\Delta U_{Cp}(mT)$ in the vicinity of the time moment $(m-1)T + \tau$, i.e. before the moment of time mT, in the neighborhood of which the adjustable edge of the control pulse of the power switch is formed. Accordingly, the calculation of the dynamic component of the input signal of the pulse-width modulator in accordance with (2) and (3) can also be performed in the vicinity of the time instants $(m-1)T + \tau$.

4 Practical implementation

Figure 2 shows the block diagram of the control device which implements the discrete law (1) of the input PWM signal.

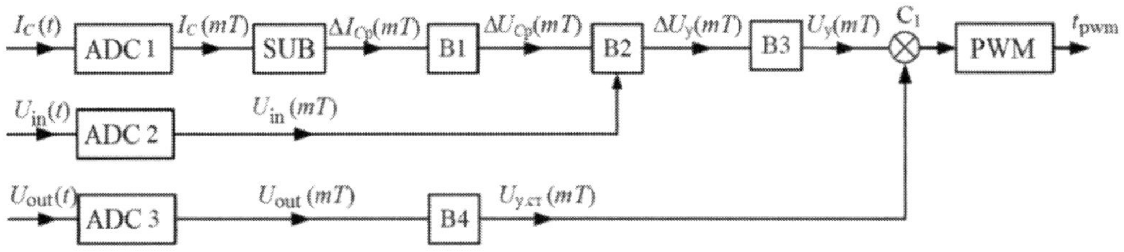

Fig. 2. Block diagram of microprocessor control device

Analog-to-digital converters ADC1 ... ADC3 provide sampling of input signals at time moments $mT+t_{OFF}$, where t_{OFF} is the offset interval and digitization of selected signal values at subsequent time intervals with a duration in the period of sampling T. The first substrator provides (5), and calculators $B1$, $B2$ and $B3$ perform calculations according to (6), (2) and (3).

5 Conclusion

The considered variant of realization (1) with application of discrete values of a regulated component of an output voltage can be used to control power modules (PM), which are the part of multimodular switching voltage regulator. The control device for PM, which is a part of a multimodal switching regulator, in addition to the tasks to ensure the required quality of the output voltage in dynamic and static modes of operation, must also solve the problem of distributing the load current between the individual modules included in the regulator. As a rule, the developers seek to ensure balancing distribution of the load current between the PMs, which allows to equalize the heat dissipation in the modules. For an even distribution of the load current between the PMs in the analog realization of the control device, a negative feedback loop (FL) is often used for the inductor current. The application of this circuit forms an oblique external characteristic of the module, which helps to balance the output currents of the PM.

In the course of tests to provide the functions (logic of operation) and the requirements of operability (fault tolerance), it was established that the layout of the digital control module completely confirmed the declared technical characteristics.

The research was made to implement decree of the Government of the Russian Federation of 9 April, 2010 No. 218 and contract between ISS JSC and the Ministry of Education and Science of the Russian Federation of 01 December 2015 No. 02.G25.31.0182.

References

1. O. A. Kozhemyak, S. K. Zeman, E. V. Yaroslavtsev, EDM, 236 (2004)
2. H. Liu, C. Fu, N. Li, A. Soldatov., W. Han, MATEC Web of Conf., (2017)
3. E. Y. Burkin, O. A. Kozhemyak, Inst. Exp. Tech, **59**, 245 (2016)
4. V. D. Yurkevich, *SIBIRCON*, 124 (2017).
5. H. Bodur, A. F. Bakan, M. Baysal, El. Eng., **85 (1),** 45 (2003)
6. M. Borage, S. Tiwari, S. Kotaiah, IEEE T. on I. Electr., **52 (6)**, 1547 (2005)
7. Y. A. Shinyakov, V. D. Semenov, V. A. Kabirov, D. S. Torgaeva, M. P. Sukhorukov and R. S. Cevastyanov, *SIBIRCON*, 346 (2017)

Kinematics of the orbital movement of a digital X-ray scanner of annular pipe seams

Rostislav Vasilchenko[1], *Fedor* Simankin[1], *Grigorii* Ziyakaev[1], and *Arkadiy* Simankin[2]

[1]National Research Tomsk Polytechnic University, 634050 Tomsk, Russia
[2]Tula State University, 300012 Tula, Russia

Abstract: The features of the kinematics of devices performing orbital displacements are considered. Different versions of the arrangement of mechanisms for the orbital movement of a digital radiographic scanner are compared. The load capacity of the support flexible joint of the mechanism was estimated.

1 Introduction

In ensuring the safety of operation of dangerous objects under pressure an important role is played by measures for non-destructive testing of the components dangerous objects [1-3]. In operation of pipeline systems under pressure and transporting corrosive liquid media, an important role is played by measures for non-destructive testing of the quality of welded joints of pipeline components. The existing methods of nondestructive testing are based on the study of the internal structure of materials and substances when they are exposed to radiographic (X-rays), infrared and gamma rays, ultrasonic vibrations, radio waves, magnetic and electric fields, penetrants, etc. Among other things, the radiographic method is more universal, since it allows testing the internal structure of the material and does not depend on climatic conditions. According to the standard [4, two types of sources are used to testing the annular weld seams: with a directional and panoramic output of X-ray radiation. A directional outlet allows the pipeline to be inspected "through two pipe walls".

The scheme with a panoramic source located inside the object of testing and concentric to its central axis is simpler and more common. Panoramic X-ray machine makes it possible to scan through the film all the annular weld seams through one wall at once. Therefore, the main scope of work on quality testing of welded seams in the construction of pipelines is performed with panoramic X-ray generators with a self-propelled movement device called a "crawler". As a rule, classical (film) radiography is used with "crawlers". The essence of the method lies in the fact that an X-ray sensitive film is superimposed on the object of testing. Herewith seam material is located between the radiation source and the film. When the radiation is emitted, the film is illuminated non-uniformly, forming a shadow image of the structure of the controlled weld seam. In this way, the existing defects in welding are visible - fissures, cracks, caverns, etc. However, despite the seeming simplicity of the testing process, the features of the film development technology do not allow to evaluate quickly the quality of the receivéd shadow image. In the event of abnormal situations, the seam testing process can be delayed for a long time.

2 Features of the kinematics of the orbital movement

The ability to assess quickly the quality of the seam image and simplify the storage and processing of information on the monitored objects is provided by digital technologies. An example is the use of digital radiographic panels. In this case, in order to obtain a shadow image of the annular welded seam, it is necessary to conduct a scanning panel along the seam, i.e., the scanner movement mechanism should provide the orbital motion of the X-ray receiver relative to the monitoring object - a full revolution along the circular path, while the scanner's working surface is always directed to the center of the circle and perpendicular to its radius. To ensure a consistently high quality shadow image of the seam, the radiation receiver must perform a movement with a constant amount of distance from the radiator and from the object of the testing. To do this, the scanner must move around the object of testing, relying on the tube itself, while maintaining the scanner's constant position in the axial direction and a stable position in the radial direction with respect to the monitor object, regardless of the angle of rotation.

The sensory element is a scintillation detector. Such detectors make it possible to obtain shadow images of the internal structure of the material; however, they are very sensitive to focusing radiation. Thus, if the panoramic source is accurately positioned relative to the central axis of the pipe and the pipe meets the requirements for circularity, to ensure the required quality of the shadow image, it is sufficient to support the movable part of the scanner on the outer surface of the pipe or to a portable supporting surface specially installed on the pipe. As such a surface, a guide belt similar in design to those used for semi-automatic pipe welding by the CRC-Evans guide belts [5] can act as such a surface. In this case, the movement mechanism of the panel is pressed against the end surfaces of the belt by special rollers, one of which receives rotation from the drive with the electric motor and causes the scanner to perform orbital motion around the pipe with the object of testing [6]. However, the experience of operating such mechanisms (the semi-automatic welding head M300 or P600 of CRC-Evans [7]) indicates a high wear rate and, as a consequence, the high cost of the drive's leading rollers. In addition, belts are also susceptible to abrasion and should be replaced frequently.

There is another solution, which implements the orbital movement of the scanner around the object of testing. In this embodiment, the movable platform of the scanner is supported directly on the pipe. Herewith the platform is pressed against the pipe by a flexible joint (belt, chain, etc.). In addition to pressing, this joint serves as a guiding support element for the traction driver's unit. Any payload, such as a radiographic panel, batteries, electronic control units, etc., can be installed on the platform. In this design, the movement mechanism of the scanner can be equipped with different types of payload, which allows to take a shadow image of the weld seam with transmission through one or two walls of the pipe.

In general, to organize the orbital movement of a digital radiographic scanner, this layout looks more promising. The supporting elements do not experience increased loads; the design of such a mechanism is simpler. The mechanism does not need additional settings when installed on the pipe, except for adjusting the flexion force. The flexible joint itself is structurally simpler than the guide belt, it is more versatile when working with different nomenclature of the pipes being studied and advantageously differs in mass-dimensional characteristics. However, simplification of the construction of the support element (in our case, flexible coupling) can lead to a loss of strength characteristics, since in the process of work considerable tensile forces are applied to it.

3 Design scheme

To evaluate the effort required for flexible communication, a design scheme was developed (Figure 1). This scheme allows to take into account the features of different layout elements on the scanner platform, quickly changing their location. For the numerical estimation of the load of flexible communication, a program was created in the MathCAD software.

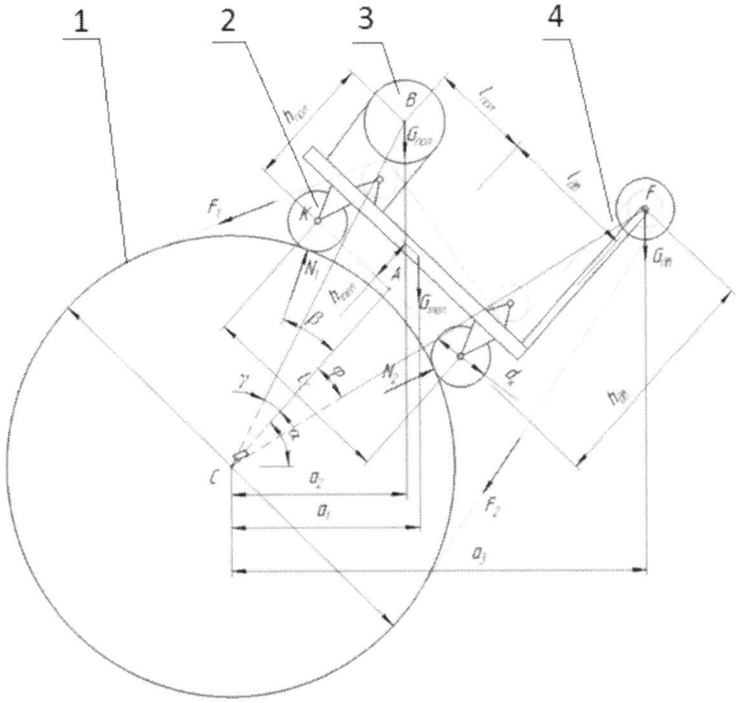

Fig. 1 Calculation scheme of the arrangement of the movement mechanism of the scanner of pipe seams: 1 - the object of testing (pipe with an annular welded seam), 2 - the platform with support wheels and the scanner drive, 3 - workload (scanner), 4 - flexible joint.

The results of the calculations (Figure 2) show forces stretching the flexible joint. The maximum forces arise on the section of the circular trajectory corresponding to the clockwise position in the range of 5 to 2 hours when moving clockwise. For example, with a payload of 40 kg, the maximum forces is a little more than 4000 N. Herewith and the torque that the scanner drive needs to keep the moving in this trajectory's section is 230 N*m. Such force values can be perceived without destruction by a flexible coupling made of metal or composite materials. To ensure flexibility, such joint must consist of separate elements that have mobility with respect to each other.

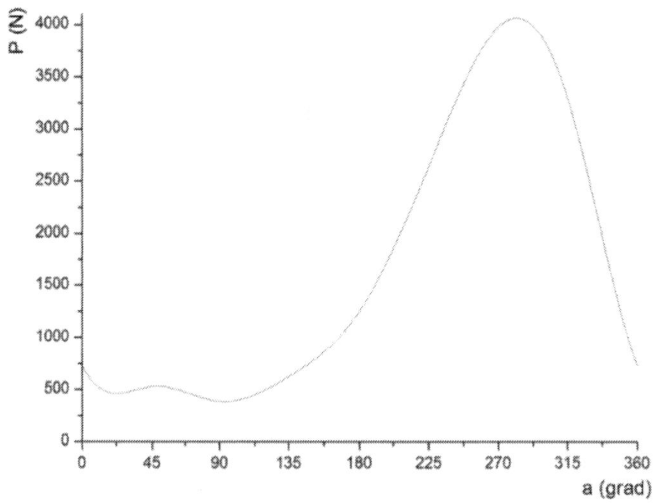

Fig. 2 The amount of tensile force applied to the flexible joint, depending on the angle of rotation of the mechanism around the pipe.

4 Conclusions

Thus, the considered arrangement of the mechanism for moving the X-ray scanner has a number of advantages over analogues using special guide belts for movement. In this complex "mechanism-object of testing" there is a flexible joint that ensures that the scanner is pressed to pipe and performs the function of a guiding support element for the traction driver's unit. To meet the conditions of strength in the process of working, this joint should be made of metallic materials. Structurally, such a joint can be made as a chain consisting of their individual strong elements, capable to swing relative to each other by a certain angle.

References

1. N. A. Ploskov, V. I. Danilov, L. B. Zuev, I. O. Bolotina, D. V. Orlova, Russian Physics Journal **54(12)** (2012)

2. V. I. Danilov, L. B Zuev, I. O. Bolotina, A. A. Zagumennyi, Ph.of M. and Metal. **106 (3)** (2008)

3. I. O. Bolotina, V. N. Borikov, V. S. Ivanova, K. V. Mertins, S. V. Uchaykin, J. of Petrol. Sc. and Eng. **161** (2018)

4. State standard 7512-82 «Non-destructive testing. Welded seams. Radiographic method»

5. Pipeline welding equipment.URL : http://www.crc-evans.com/equipment/suppliessupport/band

6. X-ray scanner for weld seams.URL : http://asgink.ru/userfiles/file/doc/Transcan-bucklet.PDF

7. Pipeline welding equipment.URL : http://www.crc-evans.com/equipment/welding-machines/p-600

Magnetic field computation and simulation of the coil systems using Comsol software

Ivan Zatonov*, *Pavel* Baranov, and *Andrey* Kolomeycev

National Research Tomsk Polytechnic University, 634050 Tomsk, Russia

Abstract. The article considers the calibration system computation with the inhomogeneity of magnetic field less than 0.1 %. Method of calculation is described. The numerical simulation using finite-element analyze was made for such system as: Helmholtz coils, improved Helmholtz coils, four and six coils system. Authors made the calculation of the magnetic field homogeneity toward axial direction. Based on procure results, authors analyzed magnetic field homogeneity and compared different types of coil systems.

1 Introduction

The measuring of magnetic flux density plays the key role in such brunches of since and technology like: task of orientation, navigation, stabilization, flaw detection, nondestructive testing, magnetic tomographic measurement, shielding action for quantum computer, etc.

Magnetometer gains currency for magnetic field measuring [1]. The stormy developing of robotic engineering complexes of surface-mounted, underwater and space system appropriation requires magnetometers evolution that will provide opportunity for weak magnetic field measuring. The theme of magnetic field measuring was reviewed by following scientists [2, 3].

Permanent field source of magnetic flux density is used for magnetometers calibration and graduation. The inaccuracy of magnetic field source must be three times less than the inaccuracy of the calibrated sensor. For instance, for the measuring device with inaccuracy near of 1 %, it is necessary to provide homogeneity magnetic field with inaccuracy 0.1-0.3 %.

Some of the most ordinary ways for homogeneity magnetic field creation is using of Helmholtz coil. There are a lot of works devoted to the calculation of Helmholtz coils magnetic field [4-8]. One of the others way for uniform magnetic field generation is to use spheroidal coils [9]. The system with bigger number of coils is used to create magnetic field with higher [10], such as: Improved Helmholtz coils, four and six coils system. In this article we made the comparison of magnetic homogeneity in axial direction for four types of coils with similar radius.

* Corresponding author: iaz5@tpu.ru

2 Finite element analysis

Authors used the program Comsol to conduct the finite element analysis. Every coil system was provided with its own task. The calculation was carried in module "Magnetic field". For the windings parameters setting the authors used Multi-turn coil section. The following parameters were set: the number of windings, the current in a conductor and the movement direction. The last parameter was set through the reference edge. It should be noted that the current movement direction in both coils match. The block with manually setting mesh was added to the model in order to receive precise measuring data avoiding fluctuation. It helped to receive data with interval lesser than 1 mm. The stationary type of the task solution was chosen for each of the models as well as iterative solver.

The calculated coils parameters values are persented in the table 1.

Table 1. Coils parameters.

i	Parameter value	
	h, m	N
Helmholtz coils		
-	0.025	10
Improved Helmholtz coils		
Central position	0	5
Edge position	0.037	10
4 coils system		
1	0.01	10
2	0.045	26
6 coils system		
1	0.009	10
2	0.029	12
3	0.0615	31
8 coils system		
1	0.0135	10
2	0.0425	12
3	0.0805	17
4	0.146	34

3 Calculation results

The Figure 1-a presents results of the magnetic field calculation of the six coils system in the axial direction with the DC value of 1 mA. The figure 1-b shows the heterogeneity for four coil system in two-dimensional space. The maximum value of the magnetic flux density heterogeneity is observed near coils surface.

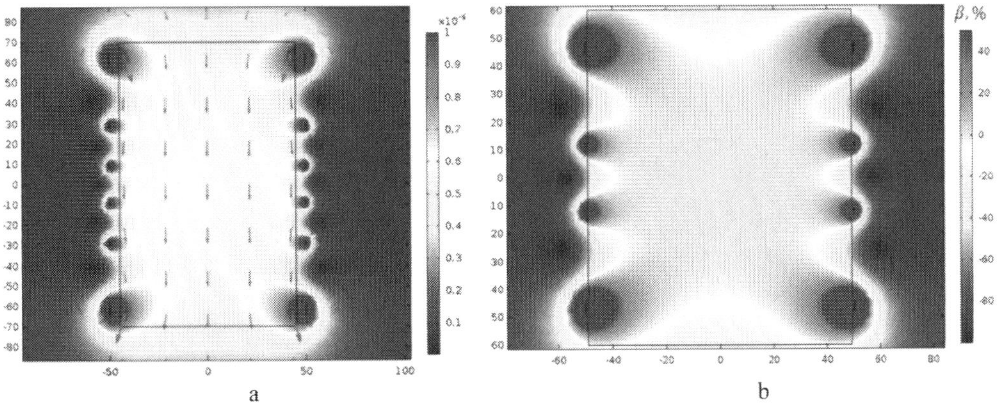

Fig. 1. a - the distribution of six coil system magnetic field, b - The heterogeneity of the magnetic field for the four coil system.

The figure 2 presents the heterogeneity of the magnetic field in the axial direction for the considered types of coil systems.

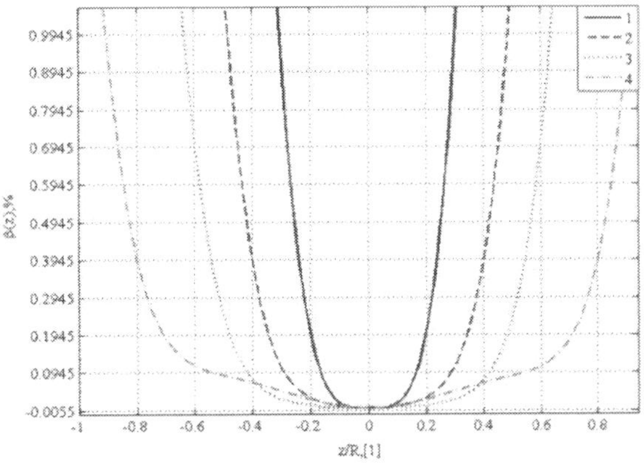

Fig. 2. The heterogeneity of the magnetic field in the axial direction, where 1 – the Helmholtz coil, 2 – Improved Helmholtz coil, 3 – four coils system, 4 – six coils system.

The changing of one of the parameters presented in the table 1 leads to the magnetic field inhomogeneity change.

The figure 3 shows that the variation of the magnetic field inhomogeneity parameter in the case of the fourth coils pairs depends on the value of 'h'.

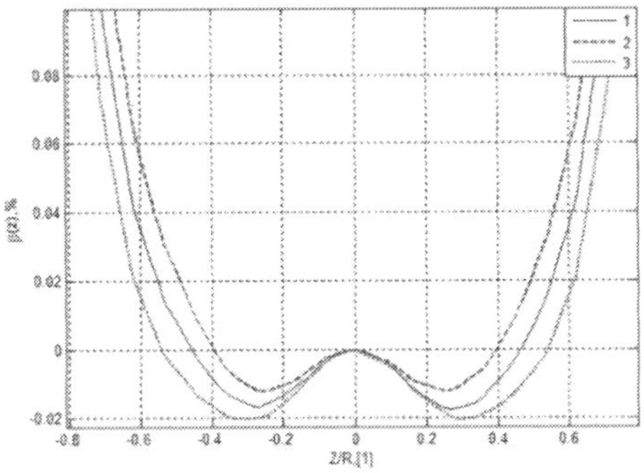

Fig. 3. The heterogeneity of the magnetic field for the 8 coils system in axial direction, where 1 – h_4 = 146 mm; 2 – h_4 = 145.5 mm; 3 – h_4 = 146.5

This figure demonstrates that a slight change in the parameters leads to increase or decrease of the field inhomogeneity. Increasing of the 'h' for the fourth pair coils allows to expand the area of the desired inhomogeneity.

Conclusion

The conducted comparison showed the needed heterogeneity of the magnetic field less than 0.1 % is provided by the four and six coils system. The increase of the coils number allows to enlarge the homogeneity of the magnetic field in the axial direction. Thus, four and six coils systems could be used to create a homogeneous magnetic field.

The research is funded from Russian Science Foundation (RSF), Grant Number 17-79-10083.

References

1. B. Ando, S. Baglio, A.R. Bulsara, C. Trigona, Sen. &Act. Phys.**151**,145 (2009)
2. T. Charubin, M. Nowicki, R. Szewczyk, Mechatron 2017 **644**,613 (2017)
3. A. Tejera,C.Carretero, J.T. Boys, IEEE Trans. Pow. Electron **32(11)**, 8349 (2016)
4. J. Wang, S. She, S. Zhang, Rev. of Scen. Instr.**75**, 2175 (2002)
5. V. Baranova, P.Baranov, Mech.&Mash.(Dyn.),1-4 (2014)
6. S. Magdaleno-Adame, JC. Olivares-Galvan, E. Campero-Littlewood, R. Escarela Perez, E. Blanco-Brisset, *In Excerpt from the proceedings of the COMSOL conference* (Comsol, Massachusetts, 2010)
7. R. Beiravand. IEEE Trans. Ind. Electron. **64**,7120 (2017)
8. M. Hosoya and E. Goto, Rev. Sci. Instrum. **62(10)**, 2472 (1991)
9. Y. Öztürk, B. Aktaş, Rev. Scien. Instr. **87**, 106103 (2016)
10. V. Ogay, Mater. Sci. Eng. **6**, 1 (2014)
11. D. Jiles, *Introduction to Magnetism and Magnetic Materials, Third Edition* (CRC Press, 2015)

AUTHOR INDEX